# SpringerBriefs in Applied Sciences and Technology

SpringerBriefs present concise summaries of cutting-edge research and practical applications across a wide spectrum of fields. Featuring compact volumes of 50 to 125 pages, the series covers a range of content from professional to academic.

Typical publications can be:

- A timely report of state-of-the art methods
- An introduction to or a manual for the application of mathematical or computer techniques
- A bridge between new research results, as published in journal articles
- A snapshot of a hot or emerging topic
- An in-depth case study
- A presentation of core concepts that students must understand in order to make independent contributions

SpringerBriefs are characterized by fast, global electronic dissemination, standard publishing contracts, standardized manuscript preparation and formatting guidelines, and expedited production schedules.

On the one hand, **SpringerBriefs in Applied Sciences and Technology** are devoted to the publication of fundamentals and applications within the different classical engineering disciplines as well as in interdisciplinary fields that recently emerged between these areas. On the other hand, as the boundary separating fundamental research and applied technology is more and more dissolving, this series is particularly open to trans-disciplinary topics between fundamental science and engineering.

Indexed by EI-Compendex, SCOPUS and Springerlink.

More information about this series at http://www.springer.com/series/8884

Thomas Brewer
Editor

# Transportation Air Pollutants

## Black Carbon and Other Emissions

 Springer

*Editor*
Thomas Brewer
Georgetown University School of Business
Washington, DC, USA

ISSN 2191-530X          ISSN 2191-5318   (electronic)
SpringerBriefs in Applied Sciences and Technology
ISBN 978-3-030-59690-3          ISBN 978-3-030-59691-0   (eBook)
https://doi.org/10.1007/978-3-030-59691-0

This Springer imprint is published by the registered company Springer Nature Switzerland AG
The registered company address is: Gewerbestrasse 11, 6330 Cham, Switzerland

# Preface

This book was created in 2019–2020 during a period of unprecedented shocks to the world economy, including all transportation sectors, as a result of the coronavirus pandemic and the onset of a severe worldwide economic recession. In the US national government, meanwhile, there was much change, confusion and uncertainty about climate change policies and related transportation and energy policies, as many of the administration's policy changes were challenged in federal courts. In Europe, BREXIT negotiations between the EU and the UK continued, and the EU-28 became the EU-27.

Throughout these tumultuous times, transportation's air polluting emissions were, nevertheless, getting serious attention from high-level officials, business executives and associations, environmental NGOs, international agencies and journalists in much of the world—and especially in Europe. This was partly because the 2019 EU elections changed the ideological composition of the European Parliament so that it became more supportive of a broad range of sustainable development policies, and the new Commission repeatedly expressed support for stronger action on climate change and other sustainable development issues. In the context of these diverse developments, a recurrent question was how the coronavirus pandemic and the enduring global recession would affect climate change policies and investments—issues that are explicitly addressed in the final chapter.

Fortunately, the experts who wrote chapters for the book persisted in their focus on the specific issues of their chapters, despite the inevitable uncertainties caused by the series of systemic shocks, and the worries, inconveniences and other circumstances affecting their professional and personal lives. I am grateful to them for their persistence. Their contributions reveal fundamental continuities in the issues addressed in the book. Most importantly, the central problem of the air pollution of the transportation sector—and its health, climate and other effects—has not disappeared. Of course, air pollution problems have been temporarily reduced by the global economic recession and the restrictions on major portions of the world economy to reduce the spread of the pandemic. Also, despite the ever-broadening range of technological applications that could be applied in all the modes of transportation, many of the familiar technological issues and the associated policy issues persist and remain highly salient.

In short, the book reflects both unprecedented recent economic changes in all regions of the world and at the same time core continuities in the nature of the problems and solutions. Although there have been numerous major changes during the past year or so, not everything has changed. The substance of the chapters, I am pleased to report, is therefore both timely and enduring.

The initial chapter introduces the types, sources and effects of transportation emissions, as well as the policies and technologies that can reduce them. The second chapter addresses the black carbon and related issues in maritime shipping, particularly in the context of policymaking at the International Maritime Organization. Three chapters then address motor vehicle issues in China, India and Europe. Following these three sector-specific chapters, the next chapter addresses transportation emission issues in all modes in the context of the European Green Deal plans and the economic recovery projects being developed in 2020. (There is no US chapter because of the unprecedented instabilities and uncertainties of US national government policies—and indeed policymaking processes and institutional relationships more generally—beginning in 2017.) The final two chapters return to more geographically diverse topics. One is a chapter on emission trading systems, including the carbon offsetting system at the International Civil Aviation Organization. The concluding chapter focuses on the implications for transportation emissions of the coronavirus pandemic and the global recession.

The idea for the project was developed by Anthony Doyle, Executive Editor for Applied Sciences at Springer's London office, in response to an article of mine on 'Black Carbon Emissions and Regulatory Policies in Transportation' in the journal *Energy Policy*. The comments of several anonymous reviewers for Springer were also instrumental in decisions regarding the focus and scope of the chapters, as Anthony and I exchanged views. On the basis of these comments and exchanges, we agreed on the subtitle, *Black Carbon and Other Emissions*. We thought that black carbon needed more attention than it had been getting, but at the same time, we were aware that there were many other air pollutants in the transportation sector. The relative importance of different kinds of emissions, however, varies among subsectors. As a result, some chapters highlight black carbon, while others address a broader array of emissions. Indeed, a central message of the collection of chapters is that there are both commonalities and differences among the subsectors. Policymakers and analysts should be attentive to both.

The production process was ably managed by Divya Meiyazhagan, who was working from India, in the midst of the extraordinarily challenging circumstances of the coronavirus pandemic in that country. I am indebted to Anthony and Divya for their professional and friendly approaches to the project.

Finally, a personal note: My preliminary work on transportation emissions began while I was a Visiting Scholar at MIT's Center for Energy and Environmental Policy Research (CEEPR), and it has continued while I have participated in a series of workshops on black carbon emissions in maritime shipping organized by the International Council on Clean Transportation (ICCT). I am indebted to both organizations for the opportunities to make presentations—and to learn from others' presentations.

Washington, USA                                                          Thomas Brewer

# Contents

# Introduction: Problems, Policies and Technologies

Thomas Brewer

**Abstract** This introductory chapter provides perspective on the global and regional contexts of the problems posed by the emissions. The focus is on climate change and public health problems—and therefore on the kinds of emissions that contribute to those problems. The chapter addresses the enduring issues: What are the emissions' chemical and physical features that are problematic? What are their effects on public health and climate change? What policies and technologies can mitigate the emissions? Black carbon (BC) receives special attention—both because it is one of the three most potent climate change forcing agents—along with carbon dioxide and methane—and because it is a worldwide public health problem causing millions of deaths per year. The coronavirus pandemic in 2020 and the global economic recession, of course, suddenly reduced the levels of emissions of all types—a topic that is considered in the final chapter. The following six chapters provide more detailed mode-specific, industry-specific and country-specific analyses of the emissions as well as mitigation policies and technologies.

## 1 Problems

Public health specialists and climate change specialists have known for many years that a wide range of transportation emissions have significant deleterious health and climate effects. Agricultural experts also recognize the problem of the effects on food production. There have accordingly been policy initiatives to reduce emissions in many countries over the years, especially with regard to emissions from motor vehicles. In recent years, there has been a more widespread recognition of the diversity of the emissions and the diversity of their health and climate effects. One aspect of the expanding scientific knowledge and the associated policy agendas is that black carbon (BC) is now more widely recognized as a major contributor to both public health and climate change problems in all regions of the world. BC has also been

T. Brewer (✉)
Georgetown University, Washington, DC, USA
e-mail: brewert@georgetown.edu

found to damage food crops. The increased knowledge and expanded policy agendas are evident in many parts of this book because the transportation sector is a major and increasing source of BC emissions in many countries. At the same time, other types of transportation air pollution are also evident in the data and on the policy agendas, and they are included in this and subsequent chapters. This introductory chapter both anticipates and reflects the contents of all of the other chapters, as it establishes the global and regional contexts for their more specifically focused analyses.

## 1.1  Types of Emissions

Fossil fuels are a major source of black carbon and other emissions, and fossil fuels are the principal fuels used in transportation worldwide. Diesel fuels have been particularly problematic sources of emissions in transportation. Table 1 lists the many types of emissions that are evident in the transportation sector, and it briefly notes salient features of each one. The list is divided into two sections: short-lived pollutants and long-lived pollutants. Short-lived climate change forcing agents have atmospheric lives as short as a week (e.g., for BC) to a few decades for others. Short-lived pollutants have been getting more attention in recent years, as there has been an intensified sense of urgency to do more to mitigate climate change forcing emissions—and to do it soon [1, 2].

Black carbon's global warming potential (GWP) at 20 years is thousands of times greater than carbon dioxide's, and at 100 years, it is hundreds of times greater [6]. Although sometimes equated with the 'soot' emitted from diesel engines and other sources, BC is actually one of several ingredients of soot. Furthermore, because it is $PM_{1.0}$—i.e., particulate matter that is less than 1 μm across and thus smaller across than a human hair—it is sometimes included in studies only implicitly as one of the constituents of $PM_{2.5}$ or in earlier public health studies as one of the constituents of $PM_{10}$. Because of their extremely small size—and thus their tendency to penetrate into lungs and blood streams—BC particles are the most potent threats among the pollutants to human health included in the studies. In recent years, public health studies [8] and climate change studies [9] and some transportation sector studies, [10–12] have explicitly focused on $PM_{1.0}$.

In its 2014 review of the literature, IPPC Working Group III concluded: 'There is strong evidence that reducing black carbon emissions from [heavy duty vehicles], off-road vehicles and ships could provide an important short-term strategy to mitigate atmospheric concentrations of positive radiative forcing pollutants' [2].

A common feature of these transportation modes' emissions—along with other motor vehicles with diesel engines—is that their emissions include both BC and organic carbon (OC). However, OC is a climate coolant. In the context of transportation emissions, this is an important fact—and sometimes a confusing one—when assessing the climate change impact of diesel engines. The ratio of BC to OC is thus a critically important metric. Since the BC/OC ratio is about 9:1 for diesel engine emissions, their emissions are clearly *net global warming agents*, and *not* net cooling

**Table 1** Transportation's air pollutants and their salient features [3–7]

| Pollutants | Salient features |
| --- | --- |
| *Short lived* | |
| Black carbon (BC) | Particulate matter ($PM_{1.0}$), not a gas; *second leading climate change forcing agent*; also causes *respiratory, cardiovascular, asthma and other diseases*; damage to plants reduces *food production*; diesel engines are a major source |
| Methane ($CH_4$) | Highly potent GHG with 20-year gloal warming potential (GWP) 84 times carbon dioxide; leaks from production, distribution and usage; atmospheric life of a decade; precursor to ozone |
| Ozone ($O_3$) | Secondary pollutant resulting from $NO_x$ and VOCs. Health effects include asthma and permanent lung damage |
| Carbon monoxide (CO) | Especially prevalent in vehicle exhaust, it causes lung disease |
| Aerosol clouds | Airplane contrails are a common form |
| Organic carbon (OC) | Climate change *coolant*; co-pollutant with BC, but in smaller volumes than BC, as emission from diesel engines |
| *Long lived* | |
| Carbon dioxide ($CO_2$) | Leading climate change forcing agent, emitted by all transportation modes' internal combustion engines |
| Nitrous oxides ($NO_x$) | Long lived with a GWP that is hundreds of times greater than $CO_2$; also a precursor to ozone |
| Sulfur oxides ($SO_x$) | Sulfur dioxide ($SO_2$) contributes directly to climate change, damages *human health* and also indirectly contributes to the formation of acid rain |
| Fluorocarbons (CFCs, HFCs, HCFCs, PFCs) | Extremely potent climate change agents, some with GWPs on the order of thousands to tens of thousands; highly variable atmospheric life times from days to thousands of years |
| Sulfur hexafluoride ($SF_6$) | Long lived (more than 3000 years) and potent (GWP more than 17,000 at 20 years and more than 23,000 at 100 years) |
| Voluntary organic compounds (VOCs) | Benzene is an example; carcinogenic health effects; precursor to ozone |

agents [13–15]. An annex to the present chapter provides additional details about the definition and measurement of BC.

Carbon dioxide is also a principal emission in the transportation sector, and it is well known as the leading climate change forcing agent in terms of its total impact over the long term. It is a long-lived pollutant with a typical atmospheric life span of about one hundred years, though some molecules last for several hundreds of years. It is used as the reference greenhouse gas (GHG) for expressing the global warming potential of other GHGs, traditionally with a 100-year time horizon, but increasingly also with a 20-year horizon. Other types of emissions also contribute to climate change and damage public health, as indicated in Table 1.

## *1.2   Levels and Trends of Transportation Emissions*

Because of the economic consequences of the coronavirus pandemic, interpretations of data must obviously be unusually cautious about recent trends and future prospects of transportation emissions. Yet, the trends through the end of 2019 were established before the pandemic became widely acknowledged. The annually based trend data available at the time of writing ended with 2019. What can be said on the basis of that data?

In the few years through 2019—just before the onset of the pandemic—transportation was the source of about one-fourth of global carbon dioxide emissions, one-sixth of all global warming gases [16] and one-fourth of black carbon emissions [17].

There have been significant variations among and within the modes of transportation. Heavy duty trucks are a particularly important source of black carbon and other emissions. Trains using electricity instead of diesel fuel are particularly low carbon-intensity mode for freight and passengers. Aviation has low black carbon emissions. International maritime emissions are significant for both merchandise trade and passenger cruise ships.

There have also been significant variations among regions, countries and areas within countries. Developing countries have generally experienced greater increases in air pollutants, including black carbon, in particular, while many developed countries have experienced declining emissions levels in some types of pollutants, including black carbon. The differences in the trends can be largely attributed to differences in regulatory policies.

There are also more specific patterns among and within countries. For instance, international maritime emissions are particularly problematic in the port areas of large cities, including cities that are home ports for cruise ships. Motor vehicle emissions are similarly especially problematic in large urban areas which are transportation hubs, and the same can be said for trains and planes. An important policy implication of these patterns is that local governments have become increasingly active in developing regulations to limit transportation emissions in all transportation modes.

These and other patterns and tendencies are discussed and documented in more detail in the following chapters.

## 1.3  Effects of Emissions

An extensive study of transportation's air polluting emissions in the form of ozone and particulate matter PM $_{2.5}$ found 361,000 premature deaths worldwide in 2010 and 385,000 in 2015, or about 11% of the global deaths from these kinds of air pollution [18]. The transportation sector modes in the study included four sub-sectors: on-road diesel, on-road non-diesel, non-road mobile and international shipping. Rail was included in non-road mobile; aviation was not included. Among the sub-sectors in the study, on-road diesel motor vehicles of all types were the principal contributor with 47% of the transportation sector's total.

The relative importance of on-road diesel motor vehicles varied across regions and countries: In Europe, on-road-diesel motor vehicles contributed 66% of the transportation share of the total within France, Germany and Italy, and they contributed 58% in other EU countries. They also contributed 66% in India. They contributed only 34% in China at that time, but it has increased since then. They contributed 43% in the USA, which has a much lower percentage of diesel-powered vehicles compared to many countries in Europe and elsewhere.

There are also consistent variations between urban and rural areas around the world. The large numbers of people in urban areas as transportation hubs almost inevitably lead to large absolute numbers of deaths. After controlling for population size, the ten cities in the world with the most deaths from transportation emissions are in Europe. These estimates are likely to be low—partly because they do not include nitrogen dioxide ($NO_2$), which is also a major transport air pollutant (see Table 1). Black carbon emissions also tend to be distributed disproportionately in major urban areas because they are hubs for ships, trains and trucks.

Black carbon is not only a major cause of deaths caused by air pollution, it is also among the three major contributors to global warming along with carbon dioxide and methane. Black carbon emissions contribute about 55% as much to global warming as carbon dioxide [9]. However, it has been relatively neglected in climate change policymaking because it is particulate matter, not a gas [19]. Measures to mitigate black carbon emissions—for instance by restrictions on motor vehicles in urban areas—have nevertheless been cited in some Paris Agreement Intended Nationally Determined Contributions (INDCs) [20]. Black carbon is also a particularly potent global warming agent in the Arctic region and is thus making a disproportionally large contribution to sea level rise and changes in Atlantic Ocean currents and hence the Gulf Stream and weather patterns in Europe [21–23].

In addition to their health and climate impacts, transportation's emissions also affect food production [24]. For instance, to the extent that maritime shipping BC emissions occur near intensively farmed coastal areas, they reduce the fields' productivity.

There are economic costs associated with all types of air polluting emissions. An estimate of the global economic cost of transportation emissions of black carbon and ozone in 2015 is 1 trillion USD (2015) [10].

# 2  Policies

## 2.1  Levels of Government

Five levels of government are actively involved in policymaking on transportation emissions of air pollutants, but the relative importance of each and the mix of combinations of them vary across the four subsectors, as indicated in Table 2. At the global level, there are specialized international agencies in the UN system for aviation and for maritime shipping—namely the International Maritime Organization (IMO) and the International Civil Aviation Organization (ICAO). There are no comparable organizations for motor vehicles or railroads.

At the international regional level, the EU is actively involved in policymaking in all four of the modes, though in different ways. In maritime shipping and aviation, the EU favors the establishment of market-based multilateral mechanisms akin to the European Emission Trading System (ETS). In the motor vehicle subsector, EU institutions are involved in establishing common EU-wide emission limits, and the EU has also used EU competition policy to fine heavy truck manufacturers for colluding to avoid adoption of more aero-dynamically configured truck cabs that would increase fuel efficiency. In the rail sector, it promotes the adoption of electric locomotives. In addition to these EU policies, there are also international regional agreements in maritime shipping in the form of emission control areas (ECAs) that include sulfur emission limits that are lower than IMO limits.

Many national governments are deeply involved in policymaking concerning all four subsectors, through both subsidies and regulations. Airports and seaports tend to be subject to local governmental authorities, though national and subnational regional agencies are also often involved. In the motor vehicle subsector, some subnational regional governments have separate emission regulations, while some local governments limit access through fee systems or outright bans of some types of vehicles. In the rail subsector, subnational regional and local governmental authorities are typically responsible for urban and regional passenger commuter systems, often with national government involvement as well. Of course, there are significant variations in these mixes of governance systems.

**Table 2** Examples of government policies at various levels

| Levels of government | Aviation | Maritime | Motor vehicles | Railroads |
|---|---|---|---|---|
| Global | ICAO | IMO | None | None |
| International regional | EU | EU; ECAs | EU | EU |
| National | Many govts | Many govts | Many govts | Many govts |
| Subnational regional | Airports | Seaports | Emission regulations | Commuter systems |
| Local | Airports | Seaports | Restricted access | Commuter and urban systems |

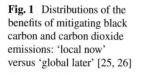

**Fig. 1** Distributions of the benefits of mitigating black carbon and carbon dioxide emissions: 'local now' versus 'global later' [25, 26]

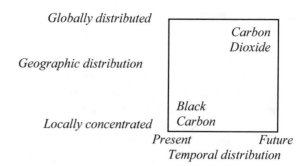

## 2.2 Two Paradigms of the Political Economy of the Effects of Mitigation Policies

A political-economy approach to the analysis of the effects of mitigating emissions focuses on identifying the distributions of the benefits and costs among groups and the distributions over time, as represented in Fig. 1.

The figure contrasts the distributions of the effects of mitigation policies of black carbon and carbon dioxide. The differences in the graph reflect fundamentally different economic and political consequences. Carbon dioxide poses the political challenge of mobilizing support for policies that promise widely distributed benefits in the distant future. Black carbon, in contrast, offers the potential of mitigation policies with localized benefits in the short term, even within days.

## 3 Technologies

Technological innovation and adoption can be retarded or induced by several kinds of government policies. Most obvious are regulations that directly limit emissions at specific levels such as those for motor vehicles' emissions of nitrogen oxides and ships' emissions of sulfur oxides. Tax policies can discourage innovation by subsidizing existing technologies, by subsidizing purchases of diesel-powered vehicles, or they can incentivize development of new technologies, for instance, by lowering taxes on electric vehicles. Other kinds of more direct subsidies favor some technologies over others, for instance, by government grants that favor sustainable fuels or propulsion systems over more polluting alternatives. A key dimension of pollution mitigation policies, therefore, is whether they directly target particular technologies for incentives or disincentives or target them in order to encourage or discourage them indirectly.

These are especially timely issues in view of the many kinds of rapidly changing technologies. Although the phrase 'technological revolution' has been overused and although advocates of some new technologies 'hype' their potentials and ignore their shortcomings, nevertheless, it is true that there are many new digital technologies now that offer interesting opportunities for reducing air polluting emissions in transportation [27].

Technologies for reducing the emissions of diesel engines are especially promising in maritime shipping, motor vehicles and railroads. Virtually all shipping vessels in international merchandise and passenger transport have been powered by diesel engines for many years, but there are now an increasing number of exceptions. Among motor vehicles, diesel engines have been widely used for large trucks and many buses and passenger cars. In the rail subsector, many locomotives have been diesel powered, though there has been a strong trend toward electricity as an alternative, especially in Europe. Aviation is different, in as much as aviation fuel is not diesel fuel.

Because of the still common use of diesel fuel in the shipping, motor vehicle and rail industries, there are similarities across three modes in issues about the technologies that can reduce emissions. For example, diesel particulate filters have been used—even legally required—in motor vehicles for many years in many countries. As a result, there are millions of motor vehicles on the road today with diesel particulate filters that reduce black carbon and other pollutants. Although there are serious issues about compliance with the testing requirements in many countries, there is agreement that the technology is effective in motor vehicles and in locomotives.

There is a different issue about the applicability of diesel particulate filters in maritime shipping, namely the feasibility and cost of significantly increasing the scale of application of the technology. Whereas the effectiveness of existing applications in motor vehicles and locomotives is well established, the application to vessels is a scaling-up challenge. Given the size of the diesel engines on large ocean-going merchandise and passenger ships, the size of an effective diesel particulate filter poses substantially different issues about costs.

## 4   Conclusion

This chapter identifies a wide range of emissions that are evident in various combinations in the four transportation subsectors of special interest—namely motor vehicles, ships, airplanes and trains. Subsequent chapters highlight more specifically the key types of emissions in the modes, and they also highlight the technological and policy issues that are of special interest in particular transportation modes and/or particular countries. These differences stimulate analytically rich discussions of the distinctive issues for individual modes and/or countries.

There are nevertheless also some underlying issues that extend across several or even all modes and countries. One is that transportation is a major source of air pollutants that affect public health and climate change. Reducing transportation air pollutant emissions is—or should be—a high priority public policy issue. Another commonality is that there are existing technologies that can reduce the emissions, and there are many more being developed. There are also policies at many levels of governance that are already in place and reducing emissions, and many more in various stages of policymaking processes. Among the many increasingly demanding analytic challenges, one of the most important is to identify—and prescribe—which level(s) of governance do what, how and why. I trust readers find that there is progress along these many analytic dimensions in these pages.

## Appendix. Black Carbon Metrics

In Table 3, the global warming impact per tonne of black carbon is compared with a tonne of carbon dioxide. In addition to the commonly used global warming potential (GWP), the global temperature potential (GTP) is also used. GWP is the integrated value of the cumulative effects of one pulse of BC at a specific time over the indicated subsequent period of time (typically 100 years, but increasingly 20 years because of the interest of the mitigating effects of BC as a short-lived pollutant).

GWP thus cumulates the effects over a specified time period, but it does not include emissions after that period of time. GTP indicates the temperature for one selected year with no weight for years before or after the selected year. GWP is more commonly used in the climate change literature, but there are examples of the use of GTP as well.

Estimates of BC's impacts by the Inter-governmental Panel on Climate Change (IPCC) have generally become larger over time. Table 3 presents data for both 20-year and 100-year time horizons. Although 100 years have been a widely used time horizon on the grounds that the average atmospheric lifetime of carbon dioxide is about 100 years, a horizon of 20 years is often used for short-lived carbon pollutants such as BC.

The table also presents four different indicators for BC—from three separate studies, one of which includes the albedo effect of BC, which reduces snow and ice reflectivity. The most inclusive indicator, which is in the bottom row, shows an estimate of 2900 for BC's GWP relative to carbon dioxide at 20 years. Despite issues about particular metrics, the basic patterns in the table are clear: BC's impact per tonne is much greater than carbon dioxide's.

**Table 3** Global warming impact per unit of black carbon emissions compared with carbon dioxide (mean estimate and 95% confidence interval) [28, 29]

| Sources and scope of emissions data | GWP 20 years | GWP 100 years | GTP 20 years | GTP 100 years |
| --- | --- | --- | --- | --- |
| Global [9] | 3200 (270–6200) | 900 (100–1700) | 920 (95–2400) | 130 (5–340) |
| Global [30] | 1600 | 460 | 470 | 64 |
| Four regions[a] [31] | 1200 (+/–720) | 345 (+/–207) | 420 (+/–190) | 56 (+/–25) |
| BC global less radiation plus albedo [9] | 2900 (+/–1500) | 830 (+/–440) | NA | NA |

[a]Regions: East Asia, European Union + North Africa, North America, South Asia. Includes aerosol-radiation interaction

# References

1. Climate and Clean Air Coalition (CCAC), Black carbon: key messages (2016). https://www.ccacoalition.org/en/resources/factsheet-black-carbon-key-messages. Climate and Clean Air Coalition. Accessed 26 Apr 2018
2. Intergovernmental Panel on Climate Change (IPCC), *Climate Change: Mitigation of Climate Change 2014* (Cambridge University Press, Cambridge, UK, 2014)
3. T. Brewer, Arctic Black Carbon from Shipping: A Club Approach to Climate and Trade Governance. ICTSD Policy Paper (2015)
4. European Environmental Agency (EEA), Status of BC monitoring in ambient air in Europe. EEA Technical Report No. 18/2013 (2013)
5. Organisation for Economic Cooperation and Development International/Trade Forum (OECD/ITF). Transport Outlook 2019. OECD, Paris (2019)
6. Intergovernmental Panel on Climate Change (IPCC), *Climate Change 2013: The Physical Science Basis* (Cambridge University Press, Cambridge, UK, 2013)
7. Intergovernmental Panel on Climate Change (IPCC), *Climate Change 2013: The Science Basis* (Cambridge University Press, 2013)
8. S. Anenberg et al., Impacts of global, regional, and sectoral black carbon emission reductions on surface air quality and human mortality deaths. Atmos. Chem. Phys. **11**, 7253–7267 (2011)
9. T. Bond et al., Bounding the role of black carbon in the climate system: a scientific assessment. J. Geophys. Res. Atmos. **118**, 5380–5552 (2013)
10. S. Anenberg, J. Miller, D. Henze, R. Minjares, A global snapshot of the air pollution-related health impacts of transportation sector emissions in 2010 and 2015. International Council on Clean Transportation (ICCT) (2019). https://theicct.org/publications/health-impacts-transport-emissions-2010-2015. Accessed 19 May 2020
11. International Council on Clean Transportation (ICCT), 4th workshop on measuring black carbon emissions: Identifying appropriate BC measurement method(s). International Council on Clean Transportation (2017). https://theicct.org/events/4th-workshop-marine-black-carbon-emissions. Accessed 2 Apr 2018
12. International Maritime Organization (IMO), Measuring Black Carbon Emissions. International Council on Clean Transportation (2012). www.wto.org. Accessed 5 Apr 2013
13. A. Azzara, R. Minjares, D. Rutherford, Needs and opportunities to reduce black carbon emissions from maritime shipping. International Council on Clean Transportation (2015). https://theicct.org/publications/needs-and-opportunities-reduce-black-carbon-emissions-maritime-shipping. Accessed 19 May 2020
14. T.L. Brewer, Black carbon problems in transportation: Technological solutions and governmental policy solutions. Working Paper, MIT Center for Energy and Environmental Policy Research (2017). https://ceepr.mit.edu/publications/working-papers/665. Accessed 19 May 2020
15. T.L. Brewer, Black carbon emissions and regulatory policies in transportation. Energy Policy **129**, 1047–1055 (2019)
16. US Environmental Protection Agency (EPA), Global Greenhouse Gas Emissions Data (2020). https://www.epa.gov. Accessed 22 June 2020
17. Climate and Clean Air Coalition, Black Carbon (2020). https://www.ccacoalition.org. Accessed 22 June 2020
18. S. Anenberg, J. Miller, D. Henze, R. Minjares, A Global Snapshot of the Air Pollution-Related Health Impacts of Transportation Sector Emissions in 2010 and 2015 (2019). https://theicct.org. Accessed 5 Jan 2020
19. Y. Yamineva, S. Romppanen, Is law failing to address air pollution? Reflections on international and EU developments. Rev. Eur. Commun. Int. Environ. Law **26**, 189–200 (2017)
20. (UNFCCC), INDCs as communicated by Parties (2020). https://www4.unfccc.int. Accessed 25 June 2020

21. Arctic Monitoring and Assessment Programme of the Arctic Council (AMAP), *AMAP Assessment 2015: Black Carbon and Ozone as Arctic Climate Forcers* (AMAP, Oslo, Norway, 2015)
22. T. Brewer, Arctic Black Carbon from Shipping: A Club Approach to Climate and Trade Governance. Presentation, COP21 side event, Paris, Dec 2015
23. B. Comer, N. Olmer, X. Mao, B. Roy, D. Rutherford, *Prevalence of Heavy Fuel Oil and BC in Arctic Shipping, 2015–2025* (ICCT, Washington, DC, 2017). https://theicct.org. Accessed 2 Apr 2018
24. D. Shindell, et al., Simultaneously mitigating near-term climate change and improving human health and food security. Science **335**, 183–189 (2012). Accessed 10 May 2017
25. T. Brewer, *The United States in a Warming World: The Political Economy of Government* (Business and Public Responses to Climate Change, Cambridge University Press, 2014)
26. T. Brewer, *Black Carbon on the Climate Policy Agenda: A New Paradigm for Analyzing Problems and Solutions* (in progress)
27. T. Brewert, Enhancing BC Regulations with New Digital Technologies. Research note and presentation at the 6th International Council on Clean Transportation (ICCT) Workshop on Marine Black Carbon Emissions, Helsinki, Finland, 18–19 Sept 2019
28. Intergovernmental Panel on Climate Change (IPCC), *Climate Change 2013: The Physical Science Basis* (Cambridge University Press, 2013)
29. T. Brewer, Arctic Black Carbon from Shipping: A Club Approach to Climate and Trade Governance, ICTSD Policy Paper (2015)
30. M. Collins, et al., Long-term climate change: projections, commitments and irreversibility. In Intergovernmental Panel on Climate Change (IPCC) (2013) *Climate Change 2013: The Physical Science Basis* (Cambridge University Press, 2013)
31. Fuglestvedt et al., Transport impacts on atmosphere and climate: Metrics. Atmos. Environ. **44**, 4648–4677 (2010)

# Maritime Shipping: Black Carbon Issues at the International Maritime Organization

Bryan Comer

**Abstract** Ships are an important and growing source of anthropogenic black carbon emissions. Black carbon emissions from ships have grown 12% between 2012 and 2018 and represent about one-fifth of shipping's carbon dioxide equivalent emissions every year, based on their 20-year global warming potential. The International Maritime Organization (IMO) has spent more than a decade studying how to define, measure, and control black carbon emissions from ships, with a particular focus on reducing the impact of black carbon emissions on the Arctic. At a recent workshop of the International Council on Clean Transportation (ICCT), participants agreed that there were six appropriate black carbon control policies that the IMO could consider, including black carbon emissions limits for ships and banning the use of heavy fuel oil in the Arctic. The IMO's separate efforts to reduce and eventually eliminate greenhouse gases (GHGs) from maritime shipping will indirectly reduce black carbon emissions. Reducing and eliminating black carbon emissions from ships can mitigate the most severe impacts of climate change, especially in the Arctic, and protect human and ecosystem health. Black carbon regulations could be in effect by 2023, although this may be delayed because IMO meetings have been postponed due to the coronavirus pandemic.

## 1 Introduction

Maritime shipping emits about one billion tonnes of climate pollution each year, about 3% of the world's anthropogenic greenhouse gas (GHG) emissions, and that is only one cost of transporting more than 80% of the world's goods [1]. If treated as a country, shipping would rank sixth, emitting more than Germany. While the International Maritime Organization (IMO)—the United Nations agency responsible for regulating the international shipping industry—has set a goal of cutting GHG emissions by at least half from 2008 levels by 2050, the sector is currently on track

B. Comer (✉)
International Council On Clean Transportation (ICCT), Washington, DC, USA
e-mail: bryan.comer@theicct.org

to double its current emissions by mid-century [2]. According to a conservative estimate, air pollution emissions from ships also cost at least 60,000 premature deaths each year [3]; others have estimated the death toll at hundreds of thousands [4].

Black carbon contributes to both shipping's climate pollution and air pollution. Small, dark, black carbon particles are emitted as a consequence of burning cheap fossil fuels in engines that can be as large as a multi-story house; these engines lack basic emission control technologies such as diesel particulate filters even though they have been used in other sectors for decades. Black carbon strongly absorbs sunlight, directly heating the atmosphere. It also reduces albedo and accelerates the melting of snow and ice after it falls to the ground. When inhaled, black carbon particles and the contaminants adsorbed to them contribute to respiratory and cardiovascular disease that can result in premature death [5–7].

Ships emitted about 100,000 tonnes of black carbon in 2018, up 12% from the 89,000 tonnes emitted in 2012 [8]. Black carbon emissions represent 7–21% of total carbon dioxide equivalent emissions from ships every year, the former being its 100-year global warming potential and the latter its 20-year global warming potential [9]. Some of these emissions occur in and near the Arctic, where black carbon has up to five times stronger warming impact compared to mid-latitudes [10]. Ships emitted more than 1450 tonnes above 59° N latitude in 2015 including 193 tonnes in the IMO's narrowly defined "Arctic" region, which is shown in the blue outline in Fig. 1 [11]. Black carbon emissions in the IMO Arctic have now exceeded 300 tonnes as of 2019 [12]. Arctic shipping is increasing [13], and ships are sailing in the Arctic for longer periods as sea ice melts due to climate change [14]. The Arctic shipping season continues to open earlier and earlier as sea ice continues to dwindle due to climate change and as ice-breaking ships plow through the little ice that remains [15].

With increased traffic comes increased use and carriage of heavy fuel oil (HFO) in the Arctic. The amount of HFO used in the IMO Polar Code Arctic has increased 75% from 2015 to 2019 [12]; black carbon emissions from ships in this region increased 85% over the same period [12]. Limiting global warming to 1.5 °C requires the world's black carbon emissions to be at least 35% below 2010 levels by 2050, according to the Intergovernmental Panel on Climate Change [16]. Achieving this will require reductions across sectors, including maritime shipping. Fortunately, there are several options today that reduce black carbon emissions by over 90%, and some technologies can eliminate black carbon emissions from ships altogether.

This chapter explains the IMO's (slow) progress on regulating black carbon from ships. It also discusses the IMO's GHG strategy and its implications for black carbon emissions. The chapter ends with a summary that urges IMO member states to regulate and eliminate black carbon emissions from ships.

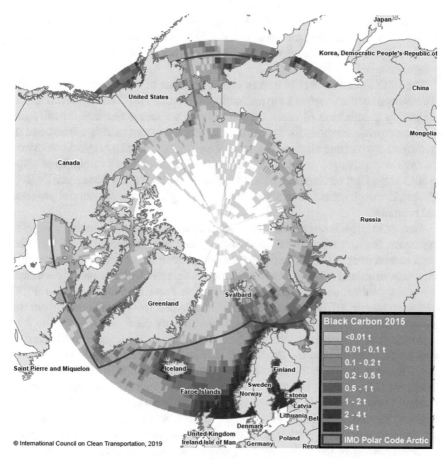

**Fig. 1** Black carbon emissions (tonnes) above 59° N latitude in 2015, including inside the "Arctic" as defined by the International Maritime Organization (waters within the blue outline) [14]

## 2 Regulating Black Carbon Emissions from International Shipping

The IMO was established in 1948 as a specialized agency under the newly founded United Nations. Safety was and continues to be one of its chief concerns. But in 1967, after the *Torrey Canyon* oil tanker shipwrecked and spilled 120,000 tonnes of oil off the coast of Cornwall, England, the IMO turned its attention towards the environment. To combat oil pollution, the IMO adopted the International Convention for the Prevention of Pollution from Ships, referred to as MARPOL, in 1973.

In 1997, MARPOL was amended to include a new section called Annex VI, which allows the IMO to regulate air emissions from ships. Regulations adopted under Annexes to MARPOL are legally binding for signatories to that Annex. Parties to Annex VI include 97 countries representing about 97% of the fleet's tonnage

[17]. Under MARPOL Annex VI, which entered into force in 2005, the IMO has regulated air pollution and carbon dioxide emissions from ships. As a consequence, black carbon emissions have also been reduced, but black carbon itself remains unregulated.

The IMO agreed in 2011 to a work plan to consider of "the impact on the Arctic of emissions of black carbon from international shipping [18]." The work was meant to develop a definition of black carbon suitable for doing research, identify the most appropriate methods for measuring black carbon from marine vessels, and to investigate appropriate black carbon control measures. Under its original work plan, the IMO was supposed to agree on appropriate actions on black carbon from ships in 2013 at the 65th session of its Marine Environment Protection Committee (MEPC 65), but the work stalled soon after it began, and by 2013, no meaningful progress had been made.

Starting in 2014, in an effort to spur progress, the ICCT began convening scientists, regulators, industry representatives, advocates, and academics at six annual marine black carbon workshops. Participants numbered around 30 each year and represented experts from around the world. Each workshop culminated with a summary document that was submitted to sessions of IMO's Pollution Prevention and Response (PPR) subcommittee or MEPC. These summaries contained recommendations from the group on how to advance the IMO's black carbon work plan.

At the first workshop, which was held in Ottawa in 2014, participants recommended that IMO accept the Bond et al. [19] scientific definition of black carbon, which it did at PPR 2 in 2015, with subsequent approval at MEPC 68 [20]. At the fourth workshop, which was held in Washington, DC, in 2017, participants recommended three methods as "appropriate" for measuring black carbon from marine engines. The IMO agreed at PPR 5 in 2018 [21]. At the fifth workshop, held in San Francisco in 2018, participants identified 13 appropriate black carbon control measures, noted in the following box, which the IMO included in its list of 41 candidate measures at PPR 6 in 2019 [22].

---

**Box 1: Appropriate black carbon control measures, as determined by participants of the ICCT's fifth black carbon workshop[a] [23]**

Fuel Type
- Liquefied natural gas (>99%)
- Distillate (33% or more)
- Biodiesel (depends on the fuel)
- Methanol (55–75% or more)

Exhaust Gas Treatment

- Diesel particulate filter alone or with selective catalytic reduction, paired with marine fuels with low sulfur and ash content such as distillates (>96%)
- Electrostatic precipitator (>91%).

Engine and Propulsion System Design

- Engine tuning to low black carbon (varies)
- Engine control technologies (varies)
- Hybrid/power storage (varies depending on how much energy the system replaces)
- Full battery-electric vessel (100%)
- Hydrogen fuel cells (100%).

Other Measures

- Shore power (100% when connected).

[a]Black carbon emission reduction potential added in parentheses by the author, based on the workshop summary document.

At the sixth workshop, held in Helsinki in September 2019, participants agreed that there were six appropriate black carbon control policies that IMO should consider, as in the following box, including black carbon emissions limits for ships; a requirement to use newer, lower-emitting ships in the Arctic; mandatory use of shore-side electricity (shore power) when ships are in port instead of idling their engines; and banning the use of HFO and switching instead to distillate fuels or other cleaner fuels.

**Box 2: Appropriate black carbon control policies, as determined by participants of the ICCT's sixth black carbon workshop[a] [24]**
Black carbon emissions limit for new ships, globally
Black carbon emissions limit for new ships, regionally
Black carbon emissions limit for all ships, regionally
Modern ship requirement
Shore power mandate
Heavy fuel oil (HFO) ban, with a switch to distillates or other cleaner fuels
[a]Order does not imply priority.

The IMO plans to make a policy decision by 2021 on how to regulate black carbon based on what it has learned from its work to date. This decision could be delayed as the organization works to establish new procedures for virtual meetings in the wake of the coronavirus pandemic. Even if delegates do agree in 2021 to regulate black carbon, under IMO rules of procedure, it will take about two additional years before any rules can be enforced. It will therefore take at least 12 years since the IMO began its work to protect the Arctic from the impacts of black carbon emissions before any international regulations can be enforced.

The IMO is currently accepting proposals for how to regulate black carbon from ships. One or more of the policy alternatives, as presented in the preceding box, will be considered. Indeed, a switch from HFO to distillate fuels for ships operating in the

Arctic was proposed by a group of environmental non-governmental organizations in February 2020 at PPR 7, which included the Clean Shipping Coalition, Friends of the Earth International, Pacific Environment, and World Wildlife Fund [25]. (These groups have "consultative status" at the IMO and can submit policy proposals and engage in discussions but do not have voting rights.) Using distillate fuels is expected to immediately reduce black carbon emissions by at least 33% compared to using "residual fuels" such as HFO [26]. Switching to distillates is also a necessary step to using diesel particulate filters that can reduce black carbon emissions by more than 90% and typically 97% or more [24].

Some will argue that the IMO has already de facto agreed to ban HFO in the Arctic because at PPR 7, the IMO agreed to draft regulatory text for the ban. The ban is primarily meant to protect the Arctic against an HFO spill but would also reduce black carbon emissions by decreasing the use of HFO, which emits more black carbon than other marine fuels. The ICCT has criticized the current ban language as inadequate, pointing to delays, exemptions, and waivers that undermine its effectiveness [27]. The first phase of the HFO ban is meant to enter into force in July 2024, but the current text exempts ships built after August 2010 whose fuel tanks are separated from the hull by at least 76 cm, which is meant to reduce the risk of fuel spills in the case of an accident. These ships can keep using HFO until July 1, 2029, under the current text. If the HFO ban had been implemented in 2019, this exemption alone would have excluded about 42% of HFO used in the Arctic [12]. An additional 43% would have been eligible for waivers, meaning that had the ban been in place in 2019, it could have only guaranteed that 16% of HFO use were banned [12]. The amount of HFO use waived and exempted is expected to grow as Arctic shipping increases and as younger ships that comply with the fuel tank regulations cycle into the Arctic shipping fleet.

IMO delegations will be able to submit other proposals to regulate black carbon to PPR 8, currently scheduled for early 2021. We expect several policy proposals to be put forward, including banning HFO in the Arctic and perhaps plans to develop a black carbon emissions standard for ships. In the short term, black carbon could be immediately reduced by prohibiting the use of HFO in favor of other marine fuels, especially distillates. In the longer term, a black carbon emission standard can secure more substantial reductions, especially if it is stringent enough to drive the use of diesel particulate filters.

## 3   IMO's Initial Greenhouse Gas Strategy and Its Implications for Black Carbon Emissions

The Paris Agreement does not include specific targets for the maritime shipping sector and, instead, left it to the IMO to devise a plan for decarbonizing international maritime shipping. Agreed to in April 2018, the IMO's initial GHG strategy aims to peak GHG emissions as soon as possible, to reduce the carbon intensity of international shipping by at least 40% by 2030 compared to 2008, to reduce absolute GHG emissions from 2008 levels by at least 50% by 2050, and to eliminate GHGs from the sector as soon as possible, consistent with the Paris Agreement temperature goals [28].

The IMO will issue a revised strategy in 2023, which could include stronger 2030 and 2050 targets. Achieving the goals of the strategy will require reducing GHGs from new and existing ships and developing new fuels and propulsion technologies that eventually result in a global fleet of zero-emission vessels. Regulations that the IMO adopts to fulfill its climate strategy might reduce black carbon emissions, but care must be taken to ensure that ships use fuels and technologies that result in low or zero-life cycle GHG emissions.

## 3.1 Reducing Emissions from New Ships

To reduce the carbon intensity of new ships, the IMO is working to expand its Energy Efficiency Design Index (EEDI) regulations. Adopted in 2011 and enforced from 2013, the IMO's EEDI regulation requires new ships to be built more efficiently over time. The EEDI is calculated as the amount of $CO_2$ emitted by a ship divided by its cargo capacity and then divided by its optimal sailing speed. Basically, it is only a proxy for a ship's carbon intensity. When first adopted, ships for which the regulation applied were required to be 10% more efficient beginning in 2015 (Phase 1), 20% more efficient by 2020 (Phase 2), and 30% more efficient starting in 2025 (Phase 3), compared to a baseline of older ships of the same size and type. In 2019, the IMO strengthened its Phase 3 targets by moving them up to 2022 for container ships, general cargo ships, cruise ships, and gas tankers, but not oil tankers or bulk carriers. Container ships also now have to be up to 50% more efficient starting in 2022, depending on their size [29].

The problem with the EEDI is that, historically, it has been too weak to encourage innovation. Shipowners have complied by building larger ships, building ships with smaller engines, or both. Additionally, because the EEDI only regulates carbon dioxide emissions, some shipowners have switched away from conventional oil-based "bunker" fuels to liquefied natural gas (LNG). LNG is mostly methane, a potent climate pollutant that traps 86-times more heat than carbon dioxide during the first 20 years after it is emitted. While using LNG emits about 25% fewer carbon dioxide emissions than conventional marine fuels, using it may actually worsen shipping's climate impacts. The ICCT found that the most popular marine LNG engine emitted 70–82% more life cycle GHG emissions than the cleanest conventional fuel [30]. The IMO regulates sulfur oxides, nitrogen oxides, and carbon dioxide, but not methane. LNG contains virtually no sulfur and burning LNG in a low-pressure injection engine results in very low nitrogen oxide emissions and lower carbon dioxide emissions than conventional fuels, but unburned methane escapes as a consequence; we call this "methane slip." LNG engines also emit only tiny amounts of black carbon emissions. So, while LNG can dramatically reduce air pollution and black carbon emissions, because of the methane emissions both upstream and from the engine, using it does not help achieve the IMO's climate goals and is actually counterproductive.

In 2020, the IMO is scheduled to begin designing new EEDI Phase 4 standards. In a submission to MEPC 75, a group of environmental NGOs have proposed incorporating methane and all other GHGs, such as nitrous oxide, into the EEDI, starting with the Phase 4 standards [25]. Nitrous oxide is important to consider including in future phases of the EEDI because one potential zero-carbon fuel—but not zero-emission fuel—is liquid ammonia. With a chemical formula of $NH_3$, burning ammonia in an internal combustion engine will emit nitrous oxide. Nitrous oxide is a powerful GHG, with a 20-year global warming potential between 264 and 268 [31]. The NGOs also said that black carbon emissions could be incorporated into the EEDI if the IMO does not adequately regulate black carbon under its existing work plan.

EEDI Phase 4 standards could be stringent enough to encourage shipowners to invest in innovative technologies such as wind-assisted propulsion or hull air lubrication to reduce fuel consumption and emissions. Ships could also use batteries or fuel cells to replace at least some of their energy use. In a recent study, the ICCT found that wind-assisted propulsion technologies (rotor sails) and hull air lubrication could significantly reduce energy use and emissions for cargo ships [32]. As EEDI standards become more stringent, they can eventually compel low-emission and eventually zero-emission vessels, especially if the EEDI begins controlling all GHGs rather than only carbon dioxide.

## 3.2 Reducing Emissions from Existing Ships

Ships are designed to operate for 25–35 years or longer. Given the long lifetimes of ships and slow fleet turnover, the IMO must regulate not only new ships, but also existing ships if it intends to meet or exceed its climate goals. To this end, the IMO is currently grappling with how to reduce emissions from the existing fleet.

The IMO already regulates some air emissions from ships. To protect human health, the IMO regulates the sulfur content of marine fuel to reduce sulfur oxide and particulate matter pollution. On January 1, 2020, the maximum allowable fuel sulfur content outside of Emission Control Areas, which have more stringent standards, fell from 3.5% to 0.50% by mass. This move saved at least tens of thousands lives per year [4]. At the same time, it reduced the climate-cooling sulfate aerosols that ships emit and has rewarded the use of sulfur-free, but methane-emitting, LNG. This has therefore intensified the need for the IMO to take actions to reduce GHGs and short-lived climate pollutants like black carbon.

While its efforts to regulate black carbon have been ongoing for about a decade, and the IMO regulates the carbon intensity of new ships under the EEDI, the IMO is just now beginning its work to regulate GHG emissions from the existing fleet. Several proposals are currently under consideration at the IMO including policies to slow ships down, to limit engine power, and goal-based carbon- or GHG-intensity standards for older, non-EEDI ships.

### 3.2.1 Slowing Ships Down

Slowing ships down, or "slow steaming," as it is called, has the potential to immediately reduce emissions from existing ships. Because ships must push their way through water, going even a little faster requires much more energy. In fact, there is a well-established cubic relationship between ship speed and fuel consumption, whereby increasing speed by 10% increases hourly fuel consumption by 33%. By this same relationship, slowing down 10% cuts hourly fuel consumption by 27%. Even though slowing down increases the time it takes to get from Point A to Point B, a 10% speed reduction reduces route-level fuel consumption by 19%. Likewise, slowing down 20% reduces fuel consumption 49% per hour and 36% per voyage. The ICCT found that when paired with efforts to strengthen new phases of the EEDI, slowing ships down improves the chances of achieving the IMO's 2050 emissions reduction targets. Strengthening the EEDI and slowing ships down 20% results in a 65% probability of cutting international shipping emissions by at least 50% from 2008 levels by 2050 [33].

### 3.2.2 Limiting Engine Power

One idea to reduce GHG emissions is to limit how much engine power a ship can use under normal circumstances. It is a roundabout way of forcing ships to slow down instead of setting a speed limit. The international shipping sector is heterogeneous, and different kinds of ships naturally sail at different speeds. For a speed limit to be effective, it would need to be different for container ships (and maybe different for different sizes of container ships), which sail relatively fast, compared to bulk carriers, which sail more slowly. Limiting available engine power, therefore, seems like a better approach. However, engine power limitation only works to reduce fuel consumption and emissions if it actually forces ships to sail slower than they already do. Unfortunately, research from the ICCT has shown that limiting engine power is not very effective at forcing ships to slow down because many ships are already sailing much slower—and therefore using much less engine power—than they used to [34]. Ships slowed down to reduce transport supply and fuel costs in the wake of the 2008 global financial crisis, a slow-steaming phenomenon that persists for the vast majority of the fleet to this day [8]. Current proposals seek to limit engine power by about 30% and to exempt ships that already meet or exceed upcoming EEDI standards. As a consequence of slow steaming and proposed exemptions, the ICCT found that limiting the engine power by 30% translates to only a 1% reduction in fuel consumption and emissions for container ships, oil tankers, and bulk carriers [34].

### 3.2.3 Goal-Based Emissions-Intensity Standards

A straightforward way to ensure that ships actually move cargo more efficiently is to regulate the amount of climate pollution ships emit per unit of transport work. Ideally, such a standard would regulate carbon dioxide equivalent emissions that account for methane, nitrous oxide, and black carbon. IMO delegations have already proposed incorporating all climate pollutants into the EEDI [35] which would regulate the technical efficiency of new ships, but operational emissions-intensity standards could also be set for existing ships. "Operational" means that the standard would apply to emissions under actual, real-world operating conditions, rather than the ship being certified under ideal test conditions, which is done for the EEDI. The benefit of such a standard is that it directly regulates the target: climate pollution emissions. It ensures that ships actually emit fewer climate pollutants per tonne-nautical mile. The challenge proponents face is that the IMO is more familiar and comfortable with regulating the technical efficiency of ships, e.g., through the EEDI, rather than operational efficiency. That is one reason why the engine power limitation proposal has gained some traction. It makes certifying compliance easier if the ship can be tested and issued a certificate every few years rather than constantly monitoring, reporting, and verifying real-world emissions. But as we have seen in other sectors, such as diesel passenger vehicles, compliance under a test procedure does not guarantee real-world emissions reductions.

All of these proposals have the potential to reduce black carbon emissions by reducing fuel consumption. However, IMO delegates must consider the life cycle emissions implications of the policy or policies they adopt to curb emissions from existing ships. For example, if biofuels are used in ships, a full life cycle accounting of their emissions will be needed. To be sustainable, biofuels would need to be sourced from wastes. These will be in short supply globally, and other sectors will be willing to pay higher prices [36]. Plus, burning biofuels will still yield black carbon emissions, which would then need to be controlled.

Speed limits would immediately reduce all emissions from ships but will not by themselves be enough to achieve zero-emissions vessels. Limiting engine power may reduce emissions but only if the engine power limit is set low enough to force ships to slow down. Emissions-intensity standards for existing ships could work, but only if they regulate more than just carbon dioxide, it would be necessary to consider the full life cycle emissions of marine fuels and or energy sources if the IMO is to do its part to decarbonize the global economy in a way that is consistent with the Paris Agreement temperature goals.

## 4 Summary

Ships are an important and growing source of anthropogenic black carbon emissions. The IMO has spent more than a decade in an effort to eventually regulate black carbon emissions from ships. The first black carbon regulations could enter into force as soon as 2023, although this could be delayed as IMO meetings have been postponed due to the coronavirus pandemic.

Regulations may focus on reducing emissions in the IMO's narrowly defined Arctic region but could expand from there. Banning the use of HFO in the Arctic is one way to immediately reduce black carbon emissions and would also enable the use of particulate filters that remove more than 90% of black carbon from the exhaust. Using other fuels, such as LNG, results in very low black carbon emissions, but increases emissions of another other climate-warming super-pollutant: methane.

The IMO's efforts to reduce and eventually eliminate GHGs from maritime shipping will also reduce black carbon emissions, but IMO delegates must take care to promote a shift towards low- and zero-life cycle emission fuels and energy sources. Wind-assisted propulsion and drag reduction technologies directly reduce all air emissions from ships, including black carbon. Batteries charged with renewable electricity as well as fuel cells powered by hydrogen produced from renewable energy such as wind, waves, and solar, would result in zero-emissions of all pollutants from the ship, as well as practically zero-life cycle emissions.

Eliminating emissions from a sector as large and diverse as maritime shipping is challenging, but it must be done. Otherwise, maritime shipping will be one of the last sectors contributing to the climate crisis. Assuming the IMO agrees to regulate black carbon, this process will have taken more than a decade; we cannot take the same plodding pace towards zero-emission vessels. A zero-emission shipping sector means not only an end to black carbon emissions from ships, but also an end to ships' contributions to air pollution, acid rain, eutrophication, and ocean acidification.

# References

1. United Nations Conference on Trade and Development (UNCTC), Review of maritime transport 2017 (2017). https://unctad.org/en
2. D. Rutherford, B. Comer, The International Maritime Organization's initial greenhouse gas strategy. International Council on Clean Transportation (2018). https://theicct.org/publications
3. D. Rutherford, Silent but deadly: the case of shipping emissions (2019). https://theicct.org/blog/staff/silent-deadly-case-shipping-emissions
4. M. Sofiev et al., Cleaner fuels for ships provide public health benefits with climate tradeoffs. Nat. Commun. 9(1), 406 (2018)
5. S. Anenberg et al., The global burden of transportation tailpipe emissions on air pollution-related mortality in 2010 and 2015. Environ. Res. Lett. 14(9), 094012 (2019). https://doi.org/10.1088/1748-9326/ab35fc
6. S.C. Anenberg, J. Miller, D. Henze, R. Minjares, A global snapshot of the air pollution-related health impacts of transportation sector emissions in 2010 and 2015 (2019). https://theicct.org/publications/health-impacts-transport-emissions-2010-2015
7. S. Anenberg et al., Global air quality and health co-benefits of mitigating near-term climate change through methane and black carbon emission controls. Environ. Health Perspect. 120(6), 831–839 (2012). https://doi.org/10.1289/ehp.1104301
8. J. Faber, et al., Fourth IMO GHG Study International Maritime Organization (2020).
9. N. Olmer, B. Comer, B. Roy, X. Mao, D. Rutherford, Greenhouse gas emissions from global shipping, 2013–2015 (2017). https://theicct.org/publications/GHG-emissions-global-shipping-2013-2015

10. M. Sand, T.K. Berntsen, Ø Seland, J.E. Kristjánsson, Arctic surface temperature change to emissions of black carbon within arctic or midlatitudes. J. Geophys. Res. Atm. **118**(14), 7788–7798 (2013). https://doi.org/10.1002/jgrd.50613

11. B. Comer, N. Olmer, X. Mao, B. Roy, D. Rutherford, Prevalence of heavy fuel oil and black carbon in Arctic shipping, 2015 to 2025 (2017). Retrieved from International Council on Clean Transportation website: https://theicct.org/publications/prevalence-heavy-fuel-oil-and-black-carbon-arctic-shipping-2015-2025.

12. B. Comer, L. Osipova, E. Georgeff, X. Mao, The International Maritime Organization's proposed Arctic heavy fuel oil ban: likely implications and opportunities for improvement (2020). https://theicct.org/publications/analysis-HFO-ban-IMO-2020

13. PAME, The increase in Arctic shipping 2013–2019: Arctic shipping status report (ASSR) #1. Arctic Council (2020)

14. B. Comer, N. Olmer, X. Mao, B. Roy, D. Rutherford, Prevalence of heavy fuel oil and black carbon in Arctic shipping, 2015 to 2025 (2017). https://theicct.org/publications/prevalence-heavy-fuel-oil-and-black-carbon-arctic-shipping-2015-2025

15. A. Adamopoulos, Russian tanker sets pace across northern sea route (2020). https://lloyds list.maritimeintelligence.informa.com/LL1132531/Russian-tanker-sets-pace-across-northern-sea-route

16. Intergovernmental Panel on Climate Change (IPCC), Summary for policymakers, in *Global Warming of 1.5°C*. An IPCC Special Report (2018). https://www.ipcc.ch/site/assets/uploads/sites/2/2019/05/SR15_SPM_version_report_HR.pdfip

17. International Maritime Organization, Status of Treaties (2020), p 3. https://www.imo.org/en/About/Conventions/StatusOfConventions/Documents/StatusOfTreaties.pdf

18. IMO Secretariat, *Report of the Marine Environment Protection Committee on its Sixty-Second Session (No. MEPC 62/24)*. International Maritime Organization (2011)

19. T. Bond et al., Bounding the role of black carbon in the climate system: a scientific assessment. J. Geophys. Res. Atmos. **118**(11), 5380–5552 (2013). https://doi.org/10.1002/jgrd.50171

20. IMO Secretariat, Report of the Marine Environment Protection Committee on its Sixty-Eighth Session (No. MEPC 68/21). International Maritime Organization (2015)

21. IMO Secretariat, Report to the Marine Environment Protection Committee (No. PPR 5/24). International Maritime Organization (2018)

22. IMO Secretariat, Report to the Marine Environment Protection Committee (No. PPR 6/20/Add.1). International Maritime Organization (2019)

23. International Council on Clean Transportation, Workshop summary: Fifth ICCT workshop on marine black carbon emissions: appropriate black carbon control measures (2018). https://theicct.org/sites/default/files/Workshop%20summary_5th%20ICCT%20BC%20workshop_vf_rev4.pdf

24. International Council on Clean Transportation, Workshop summary: sixth ICCT workshop on marine black carbon emissions: black carbon control policy (2019) https://theicct.org/sites/default/files/clean_Workshop%20summary_6th%20ICCT%20BC%20workshop_vf.pdf

25. Friends of the Earth International, World Wildlife Fund, Pacific Environment, Clean Shipping Coalition, The need for an urgent switch to distillates for ships operating in the Arctic (No. PPR 7/8/2). International Maritime Organization (2019)

26. Canada, An update to the investigation of appropriate control measures (abatement technologies) to reduce black carbon emissions from international shipping (No. PPR 5/INF.7; p. 33). International Maritime Organization (2017)

27. B. Comer, IMO's draft HFO "ban" is nothing of the sort (2020). https://theicct.org/blog/staff/imo-draft-hfo-ban-2020

28. D. Rutherford, B. Comer, The International Maritime Organization's initial greenhouse gas strategy (2018). https://theicct.org/publications/IMO-initial-GHG-strategy

29. B. Comer, Turning the ship, slowly: Progress at IMO on new ship efficiency and black carbon (2019). https://theicct.org/blog/staff/mepc74

30. N. Pavlenko, B. Comer, Y. Zhou, N. Clark, D. Rutherford, The climate implications of using LNG as a marine fuel (2020). https://theicct.org/publications/climate-impacts-LNG-marine-fuel-2020

31. Intergovernmental Panel on Climate Change (IPCC), *Climate Change 2013: The Physical Science Basis. Contribution of Working Group I to the Fifth Assessment Report of the Intergovernmental Panel on Climate Change.* (Cambridge University Press, 2013), p 1535
32. B. Comer, C. Chen, D. Stolz, D. Rutherford, Rotors and bubbles: route-based assessment of innovative technologies to reduce ship fuel consumption and emissions. https://theicct.org/publications/working-paper-imo-rotorships
33. B. Comer, C. Chen, D. Rutherford, Relating short-term measures to IMO's minimum 2050 emissions reduction target (2018). https://theicct.org/publications/short-term-measures-IMO-emissions
34. D. Rutherford, X. Mao, L. Osipova, B. Comer, Limiting engine power to reduce CO2 from existing ships (2020). https://theicct.org/publications/limiting-engine-power-reduce-co2-existing-ships-2020
35. Friends of the Earth International, Greenpeace International, World Wildlife Fund, Pacific Environment, Clean Shipping Coalition, Proposal to include all greenhouse gases emitted from ships, including methane, in the EEDI (No. MEPC 75/7/10). International Maritime Organization, (2020)
36. S. Searle, Bioenergy can solve some of our climate problems, but not all of them at once (2018). https://theicct.org/blog/staff/bioenergy-solve-some-climate-problems-not-all-once

# Transportation Air Pollution in China: The Ongoing Challenge to Achieve a 'Blue Sky'

Caroline Visser and Cristian Gonzalez

**Abstract** Air pollution is one of China's most prevalent environmental issues, and it comes at an annual cost of over 1 million lives and around 0.66% of China's GDP. Economic growth coupled with extensive motorisation has caused transport and specifically road transport to be a key contributor to air pollution through the emissions of $PM_{2.5}$, $NO_x$ and $O_3$ in particular. The Chinese government, over the last decade, has introduced an ambitious set of policies and measures to improve air quality, increasingly integrating its management with climate change management. This chapter provides an overview of general air quality policy frameworks and the measures that are specifically directed at reducing the transport impact. These include national emission standards, fuel consumption targets and fuel quality standards and the electrification of the vehicle fleet, for both private and public transport. Assessing the impact of these individual measures is difficult due to the multiple factors that determine air quality, although generally emissions of some air pollutants have decreased significantly in recent years—but others persisted or became worse.

## 1 Introduction

### 1.1 Air Pollution: The Current Situation in China

China has undergone a massive economic growth in recent decades, experiencing an annual rate of growth for the last 28 years between 6.6 and 14.2% [1]. Along with it, energy consumption increased drastically in the last 20 years by 230% [2], leading to increasing emissions and air pollution. Figure 1 illustrates the increase in $CO_2$ emissions emanating from different sub-sectors, linking it with the accumulative

C. Visser (✉) · C. Gonzalez
Global Road Links, Geneva, Switzerland
e-mail: Caroline.visser@globalroadlinks.com

C. Gonzalez
e-mail: Cristian.gonzalez@globalroadlinks.com

© The Author(s), under exclusive license to Springer Nature Switzerland AG 2021
T. Brewer (ed.), *Transportation Air Pollutants*,
SpringerBriefs in Applied Sciences and Technology,
https://doi.org/10.1007/978-3-030-59691-0_3

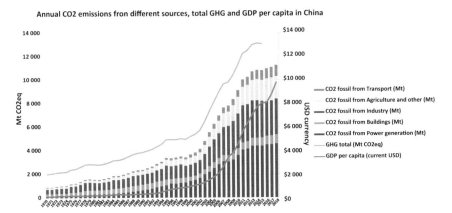

**Fig. 1** Annual China $CO_2$ emissions from different sub-sectors, total GHG and GDP per capita [3]

greenhouse gases (GHG) and GDP per capita growth over the period 1970–2018. The percentage of $CO_2$ emissions from transport in the total $CO_2$ emissions varies from 4.2 to 5.7% between 1970 and 1999 and 6.3 to 7.7% between 2000 and 2015.

The explosive growth has neither considered the effective protection of the environment nor the mitigation of the air pollution, which by now is one of the most prevalent environmental problems in China. It is brought about by multiple activities such as power generation (to a large extent through coal-fired plants), industrial activities, transport, dust, deforestation, household and commercial use of energy, waste burning and agricultural practices. Furthermore, natural factors can influence air pollution, such as geographic location, seasons, meteorological conditions and airflow patterns.

The greatest risk to human health is associated with particulate matter that is 2.5 micrometres or less in diameter ($PM_{2.5}$). This includes black carbon from combustion. Some particles are emitted directly into the air, and others are formed through reactions between gases—sulphur dioxide ($SO_2$), nitrogen dioxide ($NO_2$) and ammonia ($NH_3$)—and particles in the atmosphere. Their sources may range from power plants, in particular coal-fired ones, vehicle tailpipes, agricultural practices and wood burning to forest fires, volcanic eruptions and dust storms.

From a sample of 312 Chinese cities covered in the World Health Organisation (WHO) 2018 ambient air quality database, the average concentration of the $PM_{2.5}$ was 48 ug/m$^3$ (micrograms per cubic metre of air), more than three times that of 15 ug/m$^3$ from the world average of 2288 cities [4]. Air pollution has detrimental public health impacts in China, and it is furthermore degrading soil conditions for crop production due to the emergence of acid rain. Two empirical studies estimated independently that 1142 million and 1163 million people, respectively, died prematurely in a year in China due to two air pollutants: $PM_{2.5}$ and $O_3$ [5, 6]. Annually, air pollution costs the Chinese economy 267 billion ¥ (equivalent to US$ 38 billion), representing 0.66% of Chinese GDP [5].

## 1.2 Transport's Share in Air Pollution

The high and continued economic growth experienced by China in recent decades has been coupled with a rapid expansion of transport infrastructure and with a change in people's behaviour in transport and mobility, due to economic opportunity and increasing wealth and purchasing power, leading to extensive motorisation. Furthermore, high levels of urbanisation have exacerbated the pressure on transport systems, particularly in cities.

Figure 2 illustrates the developments in urbanisation and motorisation over 1987–2019.

The transport sector in general is claimed to be the third largest contributor to $NO_x$ pollution in China, following coal-fired power plants and industry, whereas it is the second largest source of ozone ($O_3$) emissions [7].

### 1.2.1 Road Transport

Private car use/ownership has experienced phenomenal exponential growth, as demonstrated in Fig. 2, in particular since the early 2000s. Estimates of private car emissions indicate that cars produced 20% of total carbon monoxide (CO), 24% of total nitrogen oxides ($NO_x$) and 29% of total volatile organic compounds (VOCs) in China [8]. Percentages at local level could increase between 40 and 70% [9]. Fine particulate matter $PM_{2.5}$ associated with private car use is an important source of pollution in urban and river delta areas. For example, in Beijing it accounts for 22% and in the Pearl River Delta for 35.5% of the total air pollution, respectively [9].

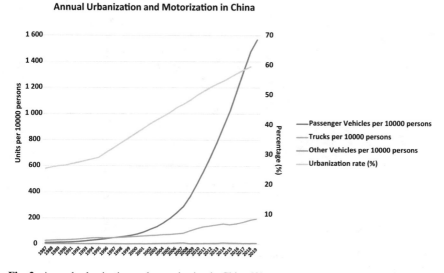

**Fig. 2** Annual urbanisation and motorisation in China [2]

According to a household survey carried out by Pew Research Centre in 2014, six out of ten households in China own a motorcycle. Motorcycles are mainly fuelled with gasoline [10]. Their contribution to CO and VOC emissions was significant at 36.6 and 68.8% in 1999 and came down to 15.7 and 25.7% in 2011, respectively [11].

Road transport has been and is increasing as the dominant mode of freight transport in China. Despite representing a modest 8% of the domestic vehicle fleet, diesel-powered heavy-duty vehicles (HDV) are claimed to be responsible for 60% of $NO_x$ and 85% of particle matter pollution on China's road networks. The average annual growth of diesel-powered HDV fleet between 2013 and 2018 was 4% [7].

### 1.2.2  Other Modes of Transport

Figure 3 depicts the overall accumulative energy consumption by the transport sector, distinguishing between the various energy sources, and the number of vehicles in use over time between 1999 and 2019. The most commonly consumed fuels during 1999–2017 in the transport sector were kerosene (+528%), diesel oil (+407%) and gasoline (+350%). Natural gas has experienced increments in energy consumption since 2008 and electricity doubled between 2010 and 2017. However, most of the electricity production is coming from coal, i.e. 65% in 2017 [12].

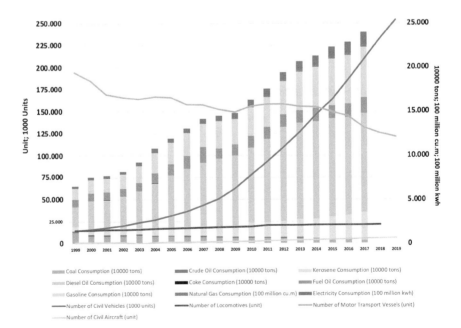

**Fig. 3** Accumulative energy consumption by the transport sector and number of vehicles in use (per transport mode) [13]

Rail and waterborne transport are generally less polluting than road transport. Rail consumes approximately 14% of energy and emits roughly 7.8% of the $NO_x$ and $PM_{2.5}$ per unit of cargo compared to road transport, whereas for waterborne transport, the percentages are 7 and 6.7%, respectively, compared to a unit of cargo transported by road [7].

For the remainder of this chapter, the main focus will therefore be on air pollution policies that are directed at reducing road transport emissions, as it is responsible for such a large share in overall transport emissions. Where specific measures regarding other modes of transport are relevant, these are briefly addressed.

# 2 National Policies and Legislation Addressing Air Pollution

## 2.1 National Policies Focusing on Air Quality

The 11th Five Year Plan for Economic and Social Development, covering the period 2006–2010, was the first FYP that contained quantitative, mandatory environment and energy targets. The 12th FYP, covering 2011–2015, targeted $SO_2$ and $NO_x$ emission specifically [14].

An "airpocalypse" smog episode during the winter of 2013, where $PM_{2.5}$ pollution caused a persistent toxic smog and low visibility over Eastern China, led to public outcry. Later in the year, the central government issued ten key actions and adopted the first Air Pollution Prevention and Control Action Plan (APPCAP). Air pollution through the emission of $PM_{10}$ and $PM_{2.5}$ particles was at the heart of the plan. It included quantitative targets for improved air quality with a time limit [15]. The key regions addressed in the first APPCAP are Beijing-Tianjin-Hebei, the Pearl River Delta and the Yangtze River Delta.

The 13th Five Year Plan (FYP) for the period 2016–2020 reinforced the government's approach by including emitted pollutants limits and enhancing the monitoring of their implementation. The plan's objective was to reduce emissions, mainly those related to industrial polluters, to promote the use of clean energy and to ensure compliance with emission standards [15]. In addition, one of the focus areas of the plan is to develop a domestic environmental technology industry that can achieve and improve the quality of the environment and ecosystems. It follows the "Made in China 2025" industrial overhaul strategy, established in 2015, that aims to provide a roadmap for the upgrade of the manufacturing sector [16].

In 2018, the second APPCAP "for winning the Blue Sky War" was launched. The three-year plan, covering the period 2018–2020, includes targets for $SO_2$, $NO_x$ and $PM_{2.5}$ emissions by 2020 and focuses more on controlling ground-level ozone ($O_3$). The plan furthermore redefined the priority geographical regions in China with heavy air pollution and replaced the Pearl River Delta with the Fen-Wei plain. The $PM_{2.5}$ targets in the second APPCAP mirror those set in the 13th FYP, i.e. a reduction of minimum 18% in $PM_{2.5}$ levels compared to 2015. However, the scope of application of the targets is widened to include all cities on prefectural or higher level that have not yet achieved them, instead of limiting it to the three priority key regions. It is observed that the $PM_{2.5}$ targets are less ambitious than some cities had set themselves in their own 13th FYP.

The 2018 APPCAP provided the first steps towards integration of air pollution and climate change management, in terms of policy and approach. Two new ministries were established: the Ministry of Ecology and the Environment (MEE) and the Ministry of Natural Resources. It meant a breakthrough in what was referred to as "bureaucratic fragmentation" [17]. Previously, the Ministry of Environmental Protection had the policy responsibility over air pollution, whereas the National Development and Reform Commission was addressing greenhouse gasses and climate change.

In 2019, the Green Travel Action Plan was issued, supported by no less than 13 ministries, including the Ministry of Transport. The plan covers plans to further stimulate the use of electric vehicles, charging infrastructure and improvements to the public transport systems to enhance green mobility [18]. The public transport improvements include the extension of inter-city high-speed rail services and subway systems and the deployment of electric busses, developments that have been set in motion in the early 2000s.

The preparations for the 14th FYP (2021–2025) are currently ongoing, and the plan is likely to be approved early 2021, after which dedicated sub-sector plans will be developed. It will be interesting to see how China will reinforce and continue its path towards tackling climate change and air pollution. Ideas on the table for the 14th FYP include a carbon emissions cap—instead of an energy consumption cap that was introduced in the 13th FYP, a further reduction of dependency on coal and targets for the percentage of non-fossil fuels (i.e. hydro, nuclear, wind and solar) in the energy mix [19]. The latter element will be of key importance to further greening the transport sector, such as decarbonising rail operations and electric vehicles through powering by clean energy sources.

## 2.2  National Legislation and Regulations

The legal basis for most policies and measures related to reducing air pollution is provided by the Environmental Protection Law, first adopted in 1989 and last revised in 2014. The law provides a framework regulating all aspects of environmental protection and pollution control. It formed the basis of an extensive set of more specific laws and measures, including the Law on Prevention and Control of Air Pollution. This "Air Law", last revised in 2015, aims at controlling air pollution, among others caused by motor vehicles and vessels. The law calls for a low-carbon, eco-friendly transportation system [15].

Equally of relevance is the Energy Conservation Law, first adopted in 1997 and last revised in 2016. The Law encourages clean and alternative fuels in transport and provides the legal basis for standards for fuel consumption limits for vehicles and vessels [15].

# 3 Measures Focusing on Reducing Transport Air Pollution

In this section, a number of measures will be discussed that have as their specific objective a reduction of pollutant emissions from transport. The first part will address national initiatives, including vehicle and fuel standards, modal shift and electrification of the vehicle fleet. The second part will address some exemplary initiatives undertaken at city level.

## 3.1 Emission Standards

### 3.1.1 Vehicle Emission Standards

The legal basis for Chinese vehicle emission standards is the "Air Law". The standards currently in force have been applied from 1 January 2018 and follow EU standards. They include the China 5 standard for light-duty vehicles, but also comprise dedicated standards for heavy-duty vehicles, three-wheeled and low-speed vehicles, motorcycles and non-road mobile machinery (NRMM) [15].

Cities and regions can, and indeed have, set their own, more stringent, standards in advance of the national implementation schedule. On a national level, implementation of China 6 standards has been delayed until January 2021, in particular the implementation of the particle number (PN) limit. However, 16 cities have already implemented the China 6 standards [20]. These can be considered to be one of the most stringent emission standards in the world.

### 3.1.2 $NO_x$ Emission Standards for Vessels

In parallel to the above-mentioned vehicle standards, the Chinese Ministry of Transport introduced stricter $NO_x$ emission standards for vessels in use for inland water transport and coastal maritime transport [21]. They apply to diesel engines on imported and Chinese flagged vessels involved in domestic transport. The vessels are to comply with the limits of the International Maritime Organisation (IMO) Tier II NOx standards.

## 3.2  Fuel Quality Standards

Fuel quality standards are issued by the Standardization Administration of China (SAC) and are setting limits to the amount of sulphur, measured in parts per million (ppm), in motor diesel and gasoline for on-road vehicles and in general diesel (for use in off-road equipment). The current standards came into force in 2017 and stipulate the nationwide availability of 50 ppm fuels by July 2017 and of 10 ppm fuels by January 2018. The standards applied in China are generally, with some exceptions, equivalent to those in use in the European Union. It is at the discretion of cities and regions, to set their own, more stringent, standards without central approval [22].

## 3.3  Fuel Consumption Standards

In its Energy Saving and New Energy Vehicle (NEV) Industry Development Plan (2012–2020), the Chinese government has set standards for different types of vehicles and vessels. Vehicles that cannot meet the standards are prohibited to be in operation. Currently in force are Phase IV standards for passenger cars, setting a fleet average target of 5.0 L/100 km for new vehicles sold in 2020 [15]. These standards include the Corporate Average Fuel Consumption (CAFC) standard for manufacturers' passenger car fleets.

The CAFC targets are embedded in a credit scheme which stipulates that car manufacturers and importers sell NEVs to earn credits amounting to a set percentage of their sales volumes of non-NEVs in a certain year. The credits differ per type of NEV, and both types of credits (NEV and CAFC) can be traded on a credit market. The scheme has come into force in 2019.

For HVDs, the current Phase III standards set limits for tractors, trucks and buses and will apply to all new vehicles in July 2021.

## 3.4  Chances of a Modal Shift?

The second APPCAP includes an ambition to increase railway frequency of passengers and freight, as well as developments and promotion of rail-water and rail-road multi-modal transport. China has heavily invested in rail infrastructure and services, mainly sub-urban rail networks and high-speed train connections. The better urban connections have to some extent led to a higher use of train services, but it appears that the shift has mostly come from people that used to take their bicycle rather than private car commuters. The construction of green zones outside mega-cities, connected with a rail transit system, is improving the quality of life of its inhabitants. However, there is no evidence this has led to decreasing motorisation levels [9].

Shifting freight transport from road to rail is not without challenges. Distorted pricing is one of the issues, which the State Council tried to address in a 2018–2020 plan. Rail has to compete with a highly competitive road freight sector and, due to the organisation of rail sector, does not seem able to respond to user requirements of speed and efficiency. Furthermore, different transport modes are not sufficiently interconnected to enable goods to be transferred between them, and rail connections to ports, logistic zones and industrial areas are inadequately developed [7].

Concerning water transport, despite having the largest system of inland waterways in the world, the lack of infrastructure efficiency, underinvestment and weak coordination and logistics at local level, the competitiveness of waterborne transport compared to rail and road is equally lagging behind [23].

With the surge in parcel freight following the uptake of online shopping for consumer goods, through platforms such as Alibaba, it can be expected that road freight will not easily be moved to more environmentally friendly but less flexible modes of transport, such as rail and water.

## 3.5 Electrifying the Vehicle Fleet

### 3.5.1 New Energy Vehicle Measures

In 2009, the Chinese government adopted a programme entitled "Ten Cities, Thousand Vehicles" to stimulate the New Energy Vehicle (NEV) market. Among the drivers to adopt the plan were to reduce air pollution, specifically in cities, and to reduce greenhouse gas emissions. Furthermore, China aimed to stimulate its domestic car manufacturing sector and to reduce its dependence on oil from the Middle East and increase energy security.

The term NEV comprises battery electric vehicles, plug-in hybrid electric vehicles and fuel-cell electric vehicles, manufactured domestically. The programme started in 2010 with trials in ten Chinese cities focusing on the roll-out of at least 1000 electric vehicles, commencing with government fleets. The programme was expanded twice to include more cities. Later in 2010, the programme was further expanded to include consumers in five cities, who were eligible to purchase subsidies for private NEV from the central government, oftentimes complemented with subsidies coming from state level [24]. Subsidy levels varied for plug-in hybrid electric vehicles (PHEVs) and battery electric vehicles (BEVs); the latter category is eligible to a higher subsidy.

Aligned with the NEV scheme, the development of infrastructure for electric car charging was accelerated, leading to a total number of charging piles of 976,00 in May 2019, including 401,000 public charging piles. 78% of the global stock of fast chargers of 150,000 units are in China [25]. Furthermore, charging facilities management systems have been improved, on both city and inter-city level. Interoperability is being achieved through the implementation of new standards for charging interfaces [26].

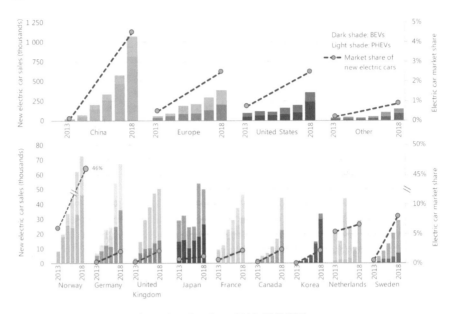

**Fig. 4** Global electric car sales and market share, 2013–2018 [27]

The subsidy scheme has stimulated the domestic production and sales of electric cars. Figure 4 depicts the sales of NEVs over the period 2013–2018 in China, compared to other key markets in the world. The total electric car fleet in 2018 in china is 2.3 million units, up 700,000 units from 2017 [25].

In April 2020, the government decided to extend the subsidy scheme up to 2022, although it will be less generous than previous versions and conditions are been tightened [28].

Electrification of medium- and heavy-duty vehicles follows a slower pace, but has equally made a start in China with estimated sales in 2018 of 1000–2000 units [27].

### 3.5.2   Greening the Public Transport Fleet

The NEV scheme also included subsidies for electric busses. China is the largest electric bus market in the world, deploying 99% of the global fleet of 460,000 units in 2018 [27]. The city of Shenzhen in Guangdong province has been championing the deployment of electric busses. In 2017, Shenzhen was the first city with a full electric public transport fleet, including 16,000 electric busses. This was achieved mainly through major government subsidies for their operation to public transport operators: subsidies of about 500,000 ¥ for each bus annually, of which 80% funded by the city of Shenzhen and 20% by the central government. The subsidy is high in relation to a purchase cost of approximately 1.8 million ¥ per electric bus. The programme in Shenzhen costs the government nearly 8 billion ¥ annually, equivalent to 1 billion US$. In 2018, Shenzhen's 13,000 fleet of taxis turned all-electric. These also receive subsidies from central and local government, albeit at a more modest level [29].

The benefits of electrification of the public transport fleet quickly manifested itself in Shenzhen; according to 2017 data of the Shenzhen Municipal Transportation Commission, $CO_2$ emissions were reduced by 1353 million tons and pollutants emissions dropped by 431.6 tons [30].

Over 30 Chinese cities are planning to fully electrify their public transport fleet by 2020. In 2018, 18% of China's bus fleet was electric [31].

## 3.6 Local Initiatives

In their attempt to ease air pollution and notorious traffic congestion, a number of Chinese city authorities have introduced demand management measures to reduce the number of cars on their networks. The most well known is Beijing's even-odd schedule, which allows even-numbered licence plates on the road one day and odd-numbered plates the next (alternating). The scheme was introduced ahead of the 2008 Summer Olympic Games and refined over the years since. The measures are enforced through traffic cameras with licence plate recognition.

Other measures that were taken to extend the benefits post-Olympics included an end-number licence plate policy—prohibiting cars with certain end number to circulate for one out of five weekdays, a yellow-label restriction policy—labelling cars that do not meet with emission standards and blocking their entering in the city and the introduction of pollution alerts with associated even-odd driving restrictions [32]. Plans for introducing London- or Stockholm-type congestion charging schemes ran into fierce opposition [33].

Restrictions similar to the end-number licence plate policy are in place in 11 other cities.

## 4  How Effective are China's Attempts at Reducing Air Pollution?

There are a number of studies that look into recent emission trends in China, which could be used as a proxy to assess the effectiveness of China's overall approach towards reducing bad air. Although they generally indicate an improvement to the air quality situation, there are many factors determining air pollution, such as meteorological conditions, that have not been taken into consideration, making it hard to attribute the effects to particular policies, especially those discussed in this chapter that addresses transport emissions specifically. Literature on this topic is scarce.

This section therefore does not aim at providing an extensive assessment of the effectiveness of those policies and measures, but describes the general trends found.

The first APPACP enabled China to make a big leap in improving air quality through setting $PM_{2.5}$ standards for key priority regions. It took drastic measures, such as closing coal-fired power generation and prohibiting coal burning for heat generation, but the city of Beijing was able to reduce annual average $PM_{2.5}$ level with 35%. All of the key regions achieved the targets. However, none of China's cities were able to achieve the WHO recommended annual average $PM_{2.5}$ levels, though, and only 107 of 338 cities of prefectural or higher level were able to reach the WHO interim $PM_{2.5}$ standard [34].

Mean $PM_{2.5}$ and $SO_2$ concentrations have declined over 2015–2017, based on measurements collected through over 1000 ground measuring stations across China. $O_3$ concentrations increased significantly, whereas mean $NO_x$ concentrations varied throughout the country and no unified trend was observed [35]. This is underwritten by data from the Centre of Research on Clean Energy, from which the below Table 1, depicting changes in pollutant levels between 2015 and 2019, is sourced [36].

**Table 1**  Changes in pollutant levels 2015–2019 in main Chinese cities and at national level [36]

|  | PM2.5 | PM10 | NO2 | SO2 | CO | Ozone |
|---|---|---|---|---|---|---|
| Anhui | -18% | -15% | 8% | -55% | -24% | 60% |
| Beijing | -48% | -35% | -27% | -67% | -46% | 1% |
| Chongqing | -30% | -32% | -12% | -53% | -24% | 20% |
| Fujian | -15% | -20% | -14% | -36% | -19% | 17% |
| Gansu | -30% | -30% | -17% | -52% | -39% | 5% |
| Guangdong | -19% | -13% | -3% | -37% | -18% | 8% |
| Guangxi | -23% | -18% | -1% | -37% | -18% | -2% |
| Guizhou | -30% | -33% | -17% | -49% | -16% | 10% |
| Hainan | -20% | -16% | -13% | -4% | -12% | 14% |
| Hebei | -34% | -30% | -16% | -62% | -33% | 19% |
| Heilongjiang | -32% | -27% | -27% | -54% | -28% | 0% |
| Henan | -23% | -26% | -16% | -70% | -38% | 23% |
| Hubei | -32% | -31% | -8% | -55% | -25% | 13% |
| Hunan | -20% | -29% | -7% | -62% | -21% | 12% |
| Jiangsu | -28% | -27% | -8% | -63% | -22% | 5% |
| Jiangxi | -20% | -18% | -2% | -55% | -15% | 24% |
| Jilin | -39% | -33% | -27% | -62% | -20% | -7% |
| Liaoning | -27% | -22% | -17% | -54% | -25% | 2% |
| Inner Mongolia | -28% | -28% | -2% | -47% | -20% | 6% |
| Ningxia | -34% | -30% | 9% | -60% | -17% | 21% |
| Qinghai | -48% | -49% | -17% | -40% | -16% | 4% |
| Shaanxi | -8% | -19% | 5% | -52% | -44% | 12% |
| Shandong | -26% | -22% | -9% | -66% | -30% | 6% |
| Shanghai | -34% | -37% | -10% | -59% | -23% | -9% |
| Shanxi | -10% | -5% | 13% | -60% | -35% | 38% |
| Sichuan | -25% | -30% | -4% | -47% | -20% | 2% |
| Tianjin | -27% | -41% | -1% | -65% | -32% | 39% |
| Xinjiang | -13% | -11% | 1% | -49% | -29% | 16% |
| Tibet | -54% | -48% | -11% | -47% | -41% | 1% |
| Yunnan | -19% | -19% | -3% | -43% | -18% | 16% |
| Zhejiang | -37% | -29% | -19% | -61% | -20% | -4% |
| National | -28% | -27% | -9% | -56% | -27% | 11% |

While the 2nd APPCAP has been underway for two years, there is no literature as of yet available that addresses its impact. Nor is there literature available that looks into the impacts of the stringent vehicle emission standards that have been put in place, at a national level, and even more stringent ones (ahead of national implementation plans) in some cities.

## 5 Concluding Observations

This chapter aimed at providing an overview of national policies and measures in place to address transport air pollution in China. In order to be effective, any policy or measure addressing transport air pollution will have to overcome the strong counterbalancing power of economic growth and commensurate increasing transport and energy demand, due to the sheer scale of the country, its economy and population. From that angle and from the perspective of the health effects, it is understandable the Chinese government refers to their attempts to curb air pollution as a "Blue Sky War". Air pollution issues differ per region in China, calling for a tailored approach. It has been observed that in the key regions Fen-Wei plain, Beijing-Tianjin-Hebei, Yangtze and Pearl River Deltas, the primary pollutants are $PM_{2.5}$ and $O_3$, whereas other regions are dealing mainly with $PM_{2.5}$ [37]. The combined effect of $PM_{2.5}$ and $O_3$ calls for more efforts towards mitigation of emissions in the transport sector.

Further decarbonising the power supply in transport, for example to electric vehicles and rail networks, is providing an opportunity to meaningfully increase the reduction of pollutants emission compared to conventional road transport.

A study into combining climate change policies with air pollution management concluded that a combination of strict (implementation of) emission standards with economy-wide pricing of $CO_2$ would deliver significant co-benefits for the reduction of air pollution, albeit mainly beneficial to non-transport sectors [38]. Further integration of air pollution and climate change management, as targeted by the institutional restructurings in 2018, might improve the effectiveness of policies.

## References

1. World Bank, World Development Indicators; DataBank (2020). https://databank.worldbank.org/reports.aspx?source=World-Development-Indicators. Accessed 15 June 2020
2. National Bureau of Statistics China, National Data (2019 and previous years). http://data.stats.gov.cn/english/Statisticaldata/AnnualData/. Accessed 14 June 2020
3. Global Road Links, Derived from data of the World Bank Development Indicators and from M. Crippa, G. Oreggioni, D. Guizzardi, M. Muntean, E. Schaaf, E. Lo Vullo, E. Solazzo, F. Monforti-Ferrario, J.G.J. Olivier, E. Vignati (2019) Fossil $CO_2$ and GHG emissions of all world countries—2019 Report, EUR 29849 EN, Luxemburg: Publications Office of the European Union. ISBN 978-92-76-11100-9. https://doi.org/10.2760/687800, JRC117610. Dataset name: EDGARv5.0_FT2018. Accessed 11 June 2020

4. WHO, WHO Global Ambient Air Quality Database (Update 2018) (2018). http://www.who.int/airpollution/data/cities/en/. Accessed 12 June 2020

5. Y. Gu, T.W. Wong, C.K. Law, G.H. Dong, K.F. Ho, Y. Yang, S.H.L. Yim, Impacts of sectoral emissions in china and the implications: air quality, public health, crop production, and economic costs. Environmental Research Letters. **13**(8), 084008 (2018)

6. Cohen et al., Estimates and 25-year trends of the global burden of disease attributable to ambient air pollution: an analysis of data from the global burden of disease study 2015. Lancet **389**, 1907–1918 (2017)

7. G. Baiyu, China's Road Freight Problem and Its Solutions| China Dialogue (2020). https://www.chinadialogue.net/article/show/single/en/11908-China-s-road-freight-problem-and-its-solutions. Accessed 12 June 2020

8. Junyu Zheng, Lijun Zhang, Wenwei Che, Zhuoyun Zheng, Shasha Yin, A highly resolved temporal and spatial air pollutant emission inventory for the pearl river delta region, china and its uncertainty assessment. Atmos. Environ. **43**(32), 5112–5122 (2009)

9. C.-L. Chen, H. Pan, Q. Shen, J.J. Want, *Handbook on Transport and Urban Transformation in China*, 1st edn (Edward Elger Publishing, 2020)

10. J. Poushter, Car, bike or motorcycle? Depends on where you live. Pew Res. Centre (2015). https://www.pewresearch.org/fact-tank/2015/04/16/car-bike-or-motorcycle-depends-on-where-you-live/. Accessed 17 June 2020

11. Jianlei Lang, Shuiyan Cheng, Ying Zhou, Yonglin Zhang, Gang Wang, Air pollutant emissions from on-road vehicles in China, 1999–2011. Sci. Total Environ. **496**, 1–10 (2014)

12. China Energy Portal. (2018). https://chinaenergyportal.org/. Accessed 16 June 2020

13. Global Road Links, Derived from data of the National Bureau of Statistics of China (2019 and previous years)

14. Yana Jin, Henrik Andersson, Shiqiu Zhang, Air pollution control policies in china: A retrospective and prospects. Int. J. Environ. Res. Public Health **13**(12), 1219 (2016)

15. L. Zhang, Regulation of air pollution. Library of Congress (2018). https://www.loc.gov/law/help/air-pollution/china.php. Accessed 12 June 2020

16. State Council of the People's Republic of China, China to invest big in 'Made in China 2025' strategy (2017). http://english.www.gov.cn/state_council/ministries/2017/10/12/content_2814 75904600274.htm. Accessed 17 June 2020

17. T. Ma, L. Qin, China Reshapes Ministries to Better Protect Environment (2018). https://www.chinadialogue.net/article/show/single/en/10502-China-reshapes-ministries-to-better-protect-environment. Accessed 12 June 2020

18. C. Nedopil Wang, Green public transport innovation in China—An opportunity for BRI countries—Green Belt and Road Initiative Center (2019). https://green-bri.org/green-public-transport-innovation-in-china-an-opportunity-for-bri-countries. Accessed 12 June 2020

19. T. Baxter, Z. Yao, The 14th Five Year Plan: What Ideas Are on the Table? Chinadialogue.Net (2019). https://www.chinadialogue.net/article/show/single/en/11434-The-14th-Five-Year-Plan-what-ideas-are-on-the-table. Accessed 14 June 2020

20. TransportPolicy.net., China: Light-Duty: Emissions| Transport Policy (2018). https://www.transportpolicy.net/standard/china-light-duty-emissions/. Accessed 12 June 2020

21. DNV-GL, New Requirements for NOx Emissions for Vessels Engaged in Chinese Domestic Trade (2018). https://www.dnvgl.com/news/new-requirements-for-nox-emissions-for-vessels-engaged-in-chinese-domestic-trade-125086. Accessed 12 June 2020

22. TransportPolicy.net., China: Fuels: Diesel and Gasoline| Transport Policy (2018). https://www.transportpolicy.net/standard/china-fuels-diesel-and-gasoline/. Accessed 12 June 2020

23. Asian Development Bank (ADB), Promoting inland waterway transport in the People's Republic of China (2016)

24. World Bank and PRTM Management Consultants, Inc., The China New Energy Vehicles Program; Challenges and Opportunities (World Bank Group, 2011)

25. International Energy Agency, Global EV Outlook 2019; Scaling-Up the Transition to Electric Mobility (2018)

26. Ministry of Ecology and Environment of the People's Republic of China, China's Policies and Actions for Addressing Climate Change (Ministry of Ecology and Environment, Beijing, 2019)
27. International Energy Agency, Global EV Outlook 2019; Scaling-up the transition to electric mobility (2019)
28. C. Randall, China Extends but Changes NEV Subsidies until 2022. Electrive.com (2020). https://www.electrive.com/2020/04/24/china-extends-nev-subsidies-til-2022/. Accessed 15 June 2020
29. D. Ren, Shenzhen's All-Electric Bus Fleet Is a World's First That Comes with Massive Government Funding. South China Morning Post (2018). https://www.scmp.com/business/china-business/article/2169709/shenzhens-all-electric-bus-fleet-worlds-first-comes-massive. Accessed 12 June 2020
30. Institute for Transportation & Development Policy, China Tackles Climate Change with Electric Buses (2018). https://www.itdp.org/2018/09/11/electric-buses-china/. Accessed 15 June 2020
31. Bloomberg New Energy Finance, Electric Buses in Cities; Driving Towards Cleaner Air and Lower $CO_2$. Bloomberg Finance L.P (2018)
32. B. Dooley, Beijing slashes traffic in pollution red alert (2015). http://news.yahoo.com/beijing-traffic-cut-schools-closed-due-pollution-red-035635673.html. Accessed 17 June 2020
33. The Economist, The Great Crawl. The Economist. 16 June (2016). https://www.economist.com/china/2016/06/16/the-great-crawl. Accessed 21 June 2020
34. F. Hao, China Releases 2020 Action Plan for Air Pollution (2018). https://www.chinadialogue.net/article/show/single/en/10711-China-releases-2-2-action-plan-for-air-pollution. Accessed 12 June 2020
35. B. Silver, C.L. Reddington, S.R. Arnold, D.V. Spracklen, Substantial changes in air pollution across China during 2015–2017. Environ. Res. Lett. **13**(11), 114012 (2018)
36. L. Myllyvirta, Air Pollution in China 2019. Centre for Research on Energy and Clean Air (2020)
37. Y. Feng, Y. Miao Ning, Y.S. Lei, Wei Liu, J. Wang, Defending Blue Sky in China: Effectiveness of the "Air Pollution Prevention and Control Action Plan" on Air Quality Improvements from 2013 to 2017. J. Environ. Manage. **252**, 109603 (2019)
38. P.N. Kishimoto, V.J. Karplus, M. Zhong, E. Saikawa, X. Zhang, X. Zhang, The impact of coordinated policies on air pollution emissions from road transportation in China. Transp. Res. Part D: Transp. Environ. **54**, 30–49 (2017)

# Road Transportation Emissions in India: Adopting a 'Hub' and 'Spoke' Approach Towards Electric-Driven Decarbonization

Mahesh Sugathan

**Abstract** India is a growing market economy, and the transport sector accounts for 12% of India's overall carbon dioxide ($CO_2$) emissions. Transport also contributes to black carbon (BC) emissions and atmospheric pollutants that are a major cause of respiratory illnesses and death. An understanding of the spatial aspects of transport emissions in India can usefully inform strategies for mitigation. The chapter examines the data around two important spatial aspects of transport emissions in India, namely cities and high-density freight corridors and some of the main policy measures deployed to address them including fuel and vehicle emission standards and modal shifts in transport. From a long-term perspective, the chapter underscores the importance of shifting towards road transport decarbonization through policies aimed at vehicle electrification and examines major policy developments and initiatives in this regard in India. It also puts forward a few considerations for policy makers with regard to deployment of electric vehicle charging infrastructure which will be a major challenge to overcome in order to scale-up manufacture and deployment of electric vehicles.

## 1 Overview of India's Transport Emissions

In 2018, India with 2.6 gigatons (Gt) of $CO_2$ accounted for 6.9% of global total emissions of $CO_2$, the fourth largest in the world after China, the USA and the European Union (EU). The year also recorded a steady growth of emissions—a 4.7% and 7.2% increase in emissions over the 2015 and 2017 levels, respectively. Over an approximately three-decade period from 1990 to 2018, India's transport $CO_2$ emissions grew by 348%, an increase second only to the power sector (where

---

The original version of this chapter was revised: The Author's e-mail address have been updated. The correction to this chapter is available at https://doi.org/10.1007/978-3-030-59691-0_9

---

M. Sugathan (✉)
Trade and Sustainability Consultant, Geneva, Switzerland
e-mail: msugathan.tradeprojects@gmail.com

T. Brewer (ed.), *Transportation Air Pollutants*,
SpringerBriefs in Applied Sciences and Technology,
https://doi.org/10.1007/978-3-030-59691-0_4

43

$CO_2$ emissions grew 462% over the same period) [1]. Transport accounts for 13% of India's $CO_2$ emissions [2]. Taking an even longer period, based on data available it is estimated that fossil fuel $CO_2$ emissions in the transport sector have increased from 50 million tonnes (Mt) in 1970 to 242 Mt in 2012, while methane ($CH_4$) and nitrous oxide ($N_2O$) emissions have increased by about 45%. There has been a decline however in the growth rates of $CH_4$ and $N_2O$ over the same period attributable mainly to the introduction of four-stroke engines in two-wheelers and three-wheelers [3]. Many studies have also attempted to estimate India's black carbon emissions with estimates of annual emissions ranging from 71.76 ~ 456 gigagrams (Gg) during the 2000s and also a wide range in terms of percentage contribution of total BC emissions by the road transport sector in India ranging from 6.5% to 34%. A more recent estimate based on isotope analysis puts black carbon emissions during the same period at 74–254 Gg/year [4].

In addition to contributing to global warming with their adverse effects on sea-level rise, weather patterns and rainfall, air pollutants from fossil fuels used in transport are also a major health hazard in the major cities of India. According to one study, out of 800,000 deaths attributable to particulate matter 2.5 (PM2.5) and ozone in India, 74,000 are attributable to transportation. Of these 66% were attributable to on-road diesel vehicles, 10% to on-road non-diesel vehicles, 19% to non-road mobile sources (including agricultural and construction equipment) and 5% to international shipping [5]. India also imports 80% of its crude oil (amounting to nearly 7% of its GDP), and more than 40% goes into the running of vehicles [6].

The energy demand within the transport sector in India is projected to be driven by the demand from the light-duty segment (largely, personal vehicles) and heavy-duty trucks. According to the International Energy Agency (IEA), the share of energy consumption by light-duty vehicles is projected to increase from 13% in 2013 to 27% by 2040 and heavy-duty trucks from 23% in 2013 to 34% in 2040 (See Fig. 1).

While increased use by heavy-duty trucks is a reflection of a growing economy and dependency on roadways-based freight in India, the increase in car share is primarily due to growing automobile dependence in cities. According to one estimate,

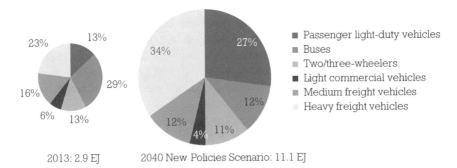

2013: 2.9 EJ          2040 New Policies Scenario: 11.1 EJ

**Fig. 1** Energy use by different transportation modes (2013 and 2040) (EJ-Exajoules where 1 exajoule = one quintillion (1018) joules) [6]

trucks and non-passenger light vehicles (i.e., those used for transporting goods) produce 47% of all $CO_2$ emissions from road transport in India [7]. Overall, heavy-duty trucking also accounts for more than half of all black carbon emissions from the transportation sector in India. In India, the average PM emissions factor for trucks is estimated to be 0.28 g/km, compared to 0.05, 0.2 and 0.03 g/km for two-wheelers, light passenger vehicles and cars/jeeps, respectively. The difference is partly explained by difference in fuel types—while trucks and non-passenger light vehicles (i.e., those used for transporting goods) typically operate on diesel fuel, two-wheelers, light passenger vehicles and cars/jeeps typically use petrol (gasoline), which produces less black carbon per kilogram combusted [7].

Consequently, from a climate-change mitigation perspective, it would be worth paying attention to two spatial aspects associated with transport-related emissions: *firstly* those generated within cities and *secondly* those generated along high-density freight corridors where freight trucks ply.

## 2 Spatial Aspects of Transport-Related Emissions in India

### 2.1 Cities

Transport-derived emissions are largely concentrated in cities and other urban centers in India. A number of the most polluted cities in the world are in India and driven by rapid urbanization which grew from 17% in 1951 to 31% in 2011 driven mainly by rural to urban migration [8]. India's urban population is expected to grow from 410 million in 2014 to 814 million by 2050. By 2025, it is projected that 46% of Indians will live in cities with more than 1 million people. Four Indian cities, Ahmedabad, Bangalore, Chennai and Hyderabad with currently 5–10 million inhabitants, are projected to become megacities in the coming years with a total of seven megacities (including Delhi, Mumbai and Kolkata projected in the country by 2030 [9]). Motorization in India is increasing at an accelerating pace. While it took 60 years (1951–2008) for India to cross the mark of 105 million registered vehicles, it took only another six years (2009–15) to add the same number. The share of public transport is expected to decrease from 75.5% in 2000–01 to 44.7% in 2030–31 with demand for personal vehicles likely to increase for commuting purposes in these cities with a resulting rise in emissions [6].

A number of policies have been introduced with a view to reducing emissions in Indian cities. Under the amended Motor Vehicles Act-2019, penalties have been further increased on vehicles not carrying a 'pollution under check' (PUC) certificate. PUC tests are mandatory every three months in Delhi and six months in other states for commercial vehicles, but implementation gaps remain [10]. Echoing the global trend, the government has also introduced stricter fuel-quality and vehicle emission standards as well as investments in expansion of mass rapid transit particularly suburban rail, light rail and metro systems. With regard to fuel-quality standards,

India is in transition to the use of ultra-low-sulfur diesel by 2020 to meet Bharat VI vehicle emission standards (BS-VI equivalent to Euro VI) announced in 2016 and implemented since 1 April 2016, thereby ensuring a leapfrog from Bharat IV (Euro IV) standards. Almost all Indian refineries have started the production of low sulfur diesel by end of 2019, and distribution had started to cities outside of the National Capital region. Indian auto manufacturers have started to roll out BS-VI compliant vehicles, and there has been a shift away from small diesel cars that have become less attractive due to the end of subsidies for diesel and higher maintenance costs of BS-VI diesel vehicles [11].

The introduction of new generation BS-VI vehicles is expected to be significantly cleaner than BS-IV with particulate matter limit being 82–93% lower and nitrogen oxide (NOx) emissions 68% lower for different segments of diesel cars. In the case of buses and trucks, particulate limits will be 50–67% lower than BS-IV levels. While maximum emissions benefits are realized from the combined introduction of BS-VI fuels and vehicle technology, a drastic cut in fuel sulfur to 10 parts per million (ppm) will also reduce particulates and sulfur dioxide emissions from all on-road vehicles to some extent. Requirements for two-wheelers will become almost as exacting as cars with separate regulation of NOx and hydrocarbon emissions compared to combined tests that existed previously [12]. A few other key features, new requirements and innovative developments associated with BS-VI standards are:

- New tests for vehicle certification that will require counting of tiny particles in the exhaust as opposed to the current practice of only weighing mass of particles. This will ensure use of the most effective diesel particulate filter, with over 95% efficiency to trap toxic particles.
- The introduction of new diesel emission control systems including advanced particulate filters for particulate control and lead NOx traps, selective catalytic reduction (SCR) and exhaust gas recirculation for NOx control.
- Onboard diagnostics (OD) system to be made mandatory for all vehicles.
- Introduction of portable emission monitors on roads to measure emissions in real-world driving conditions aligning with the European practice to prevent scandals such as Dieselgate and the one that hit Volkswagen [12].

The adoption of Euro VI equivalent standards leads to clear benefits both with regard to black carbon emissions as well as other short-lived climate pollutants by 64% by 2030 under an Accelerated Policy Scenario projection by the International Council on Clean Transportation (ICCT). This also results in significant health benefits with 1.2 million premature deaths avoided according to ICCT estimates between 2020 and 2050 with the implementation of BS-VI emission standards [13].

At the same time, because half of India's vehicular emissions are from vehicles more than 10 years old, the near-term benefits of BS-VI standards can be significantly enhanced by scrapping older vehicles, especially commercial vehicles. Further rapid expansion of the vehicle market in the coming decades means that BS-VI can only slow down but cannot stop a longer-term increasing trend in emissions as projected by the ICCT. Additional measures will be required including accelerating the retirement of BS-III and IV vehicles after 2025, looking for emission reductions beyond-and an ambitious push on electric vehicles [11]. The Union Ministry of Road Transport and Highways is amending technical regulations for the next package of reforms[1] to be introduced from 2023 onwards [12].

A major implication from the findings of a study by the Centre for Science and Environment (CSE) in India is that modal shifts towards public transport with cities and better urban design and planning, switch to clean fossil fuels such as compressed natural gas, vehicle efficiency and fuel-quality standards can only go some way towards curbing or slowing down transport-related emissions in the short to medium term. The sheer growth in urban dwellings, urban sprawl, population and vehicle stock and growth in demand for personal vehicles could overwhelm beneficial impacts of these measures in the longer term. Hence, it is imperative that thought and attention be given to a switch towards zero-emission road vehicles in all segments—two-wheelers, automobiles, buses, cargo vans and trucks. This would also need to be accompanied by investments to decarbonize electric grids supplying power to Indian mega and metropolitan cities thereby ensuring that public transit systems such as local and metro trains are also completely greenhouse gas (GHG)-free from an operational or 'wells-to wheels' perspective (i.e., emissions imputable to both the production of the fuel and its source).

## 2.2 High-Traffic Freight Corridors

The contribution of heavy-duty trucks to road transport emissions in India has been discussed earlier. The movement of heavy-duty trucks in India is mainly along high-traffic freight corridors. As of 2018, 59% of freight in India was road-dominated with 35% met by rail, 6% by waterways and only less than 1% by air. While the cost of moving freight by rail is much cheaper at Indian INR 1.41/ton-km compared to road at INR 2.58/ton, rail is economical only over longer distances on high-traffic routes and has traditionally focused on bulk goods such as coal. In addition, road transport offers greater reliability, lower transit times and end-to-end connectivity which may be crucial for time-sensitive goods. In countries sharing similar geographical and freight compositions to India such as the USA and China, truck shares are as low as 30% and 40% of total tonne-kilometers compared to 60% in India. This suggests a substantial potential for India to shift from road to rail. There is also significant need for expansion of rail freight capacity given excessive capacity utilization over existing rail corridors. The eastern freight corridor between Delhi and Howrah and western freight corridor between Delhi and Mumbai are seeing line capacity utilization between 115% and 150%. About 66% of the sections along the 'Golden Quadrilateral' and diagonal routes connecting Delhi, Mumbai, Kolkata and Chennai have

---

[1]Some of these include deciding the confirmatory factor for in-service compliance, market surveillance and independent verification testing of in-use vehicles by regulatory authorities, adoption of more stringent driving cycle for emissions testing (world not to exceed cycle), public disclosure of emissions data by the manufacturers on publicly accessible Web sites and on-board fuel consumption meters among others.

line capacity utilization of more than 100%. In the case of road traffic, the highest freight volumes (40% of the total) ply along long haul national highway corridors connecting Delhi, Kolkata, Chennai, Kochi, Mumbai and Kandla while comprising less than 0.5% of road capacity [14].

The Indian government has begun major projects to expand rail infrastructure to handle both freight and cater to the expected increase in passenger traffic between urban centers. Construction has begun on the first high-speed line between Mumbai and Ahmedabad, and additional high-speed rail corridors are planned connecting major cities and high-density passenger routes. Two dedicated freight corridors along the eastern and western freight corridor routes between Delhi and Howrah and Delhi and Mumbai totaling more than 3000 kms is scheduled for completion by 2021. Being developed by the Dedicated Freight Corridor Corporation of India Limited (DFCCIL), the US$ 12 billion project is being partly financed by the World Bank and Japan to the tune of US$ 1.86 billion and US$ 5.2 billion, respectively. Once completed, at least 70% of the freight trains will be transferred onto the DFCCIL network which will help in timely movement of cargo [15]. The routes will also be able to carry longer freight trains, with higher loads and speeds and also free up existing lines for passenger traffic at higher speeds increasing reliability.

Modal shift of freight from heavy trucks to rail (as well as passenger traffic from aviation, personal vehicles and fossil fuel-powered buses to rail) has clear environmental benefits. Presently, rail in India uses 0.04 thousand barrels per day (mb/d)of oil and 22 terawatt hours (TWh) of electricity, while at the same time avoiding 0.6 mb/d of oil use. Trains emits 29 Mt $CO_2$-equivalent (eq), while avoiding 98 Mt $CO_2$-eq and 60 kilotons (kt) of PM2.5 emissions due to displacement of traffic that would otherwise have occurred especially through fossil fuel-based road transport and aviation [16]. Further progress in decarbonization of India's electricity grid could bring down emissions even further.

## 3 Considerations and Options for Long-Term Road Transport Decarbonization in India: Adopting a 'Hub' and 'Spoke' Approach

It seems clear that higher fuel and vehicle emission standards as well as modal shift to railways will only go so far in bringing down emissions and air pollution levels. For the longer term, it is imperative that India sets as its goal a full de-carbonization of road transport. At present, electric vehicles appear to hold out the most promise for decarbonization although other technologies such as hydrogen fuel cell technology are also starting to make headway. A shift to shared, electric and connected mobility in India could help save USD 300 billion worth of oil imports and nearly a gigatonne of $CO_2$ emissions by 2030 [17]. Electric vehicle (EV) deployment is still in very early stage in India. EV sales in 2017–2018, both two- and four-wheelers, accounted for less than 0.3% of the total new vehicle sales in India [18].

**Table 1** Composition of vehicles on Indian roads based on sales figures [19]

| Vehicle type | % |
| --- | --- |
| Two-wheelers | 79 |
| Three-wheelers (passengers and goods) including tempos | 4 |
| Buses and large goods vehicles (trucks) | 3 |
| Economy four-wheel vehicles (less than INR 1 million) | 12 |
| Premium Economy four-wheel vehicles (more than INR 1 million) | 2 |

A major difference between India and most other global markets is the difference in the types of road vehicles in use. As Table 1 shows, two-wheelers are the predominant vehicle category in India while global EV technology firms are more focused on the higher end-four-wheel segment that accounts for only 2% of vehicle numbers in India. The former segment could therefore be a good opportunity for India to establish manufacturing and technological leadership while the premium segment could offer opportunities for early adoption by high-end customers and introduce best in the world technologies in India and an eco-system of component manufacturers and service providers.

More than 90% of EVs on the road are low-speed electric scooters running on lead-acid batteries. The production of low-cost three-wheeler 'e-rickshaws' has also taken off, the fastest-growing EV segment presently in India with 1.5 million e-rickshaws providing last mile connectivity in cities and smaller urban towns. This exceeds the total number of electric cars sold in China since 2011 and sales are growing at more than 20% a year since 2015. They are cheaper to run than petrol rickshaws with a cost of INR 2.1 per km compared to INR 3 for petrol rickshaws [20]. While largely driven by the unorganized sector with little government intervention, bigger players such as the Indian auto manufacturer Mahindra are also entering the market. This should lead to better quality, standardization and safety in equipment used. While some cities such as Kolkata (a mega-city) and Kanpur, Lucknow and Dehradun (smaller Tier 2 cities) have started restricting permissible routes to e-rickshaws to avoid congestion due to their slow-moving nature, they are allowed to ply on tertiary routes and outside city limits [21].

India has embarked on an ambitious plan to pursue electric mobility. A number of policy initiatives have been launched starting with the National Electric Mobility Plan 2020 (NEMP) in 2013. Subsequent initiatives and incentives are being channeled through the Faster Adoption and Manufacturing of (Hybrid &) Electric Vehicles (FAME-India) schemes. The first FAME-India scheme (FAME 1) was launched in 2015 and provided a demand incentive for consumers which was reimbursed to manufacturers. It also provided subsidies for pilots for 390 electric buses, 370 electric taxis and 720 electric autos spread across 11 Indian cities under the scheme. Phase II of FAME was launched in March 2019 with a bigger budget of Rupees 10 crores (INR 100 billion or over USD 1.3 billion) for demand incentives provided according to battery size of the vehicle and only to advanced battery chemistries such as lithium-ion. It applies mainly to public transport or commercial vehicles in three-wheeler, four-wheeler and bus segments while not excluding privately owned two-wheelers. In 2019, the Ministry of New and Renewable Energy (MNRE) set up an expert committee to propose a draft for setting up a National Energy Storage Mission (NESM), by creating an enabling policy and regulatory framework that encourages manufacturing, deployment, innovation and further cost reduction. While the central governments' policies have provided a guiding framework, many state governments, who will be the ones implementing EV-related policies and measures on ground and in major cities, have also announced their own EV-related policies and targets [22].

Detailed techno-economic scoping and feasibility exercises for EV deployment and manufacture as well as stakeholder consultations with various ministries, state and city officials, industry associations, manufacturers and consumer groups have also been undertaken by the NITI Aayog—the Indian government's policy think-tank, the Energy and Resources Institute (TERI) as well as the Bureau of Energy Efficiency and Ministry of Power and private sector consultancies such as Ernst and Young. It is beyond the scope of this article to analyze or summarize all of the findings, but in a notable aspect, various fiscal and non-fiscal incentives have been proposed, some of which have already been adopted. These include reductions in goods and services tax (GST), import-duty waivers on EV components for manufactures, waiver of registration charges for EVs accompanied by an increase in registration charges on conventional vehicles, discounts on third-party insurance and preferential rates on EV charging announced by several State Electricity Commissions [23]. Other proposals include free-parking, toll exemptions and income tax exemptions and subsidized access to land for charging infrastructure providers [24].

Initial EV support has focused on two-wheelers as well as public vehicle fleets such as shared taxis and transport buses. A number of Indian companies such as Mahindra and Tata, foreign firms such as Hyundai and a large number of smaller companies are now active in the EV mobility space even starting to entering niche segments such as trucks. Further reductions in battery costs would further boost cost-competitive manufacturing.

Scale-up of EVs will depend on the effective deployment of charging infrastructure based on a sustainable model. There were approximately 250 eV charging stations across India as of 2019, out of which major deployments are in Delhi and Bangalore. A number of recommendations have been made in this regard including the need for technology-agnostic charging, addressing standardization issues, electricity pricing and role of cities and governments in providing low-cost access to land in strategic points where the role of city authorities would be particularly important.

The need for viability gap funding in order to overcome initial low levels of demand has also been made [24]. It is also recognized that the automobile industry, saddled with costs of upgrading to new fuel and vehicle efficiency standards, will need breathing time before an all-out transition towards electric vehicles. FAME-2 sets an indicative target of 2700 charging stations in cities above 4 million inhabitants (as part of a first phase of priority roll-out within 1–3 years), bigger cities such as state capitals (as part of a second phase within 3–5 years) and fast-charging stations along major highways at an interval of about 25 km each and ultra-fast-charging stations every 100 km. The goal is to have at least one charging station in a grid of 3 × 3 kms [25]. India also updated its Model Building By-Laws from 2016 to mandate 20% of parking space within residential and non-residential complexes that must also provide EV charging infrastructure[26] and placed a cap on the maximum tariff that can be asked by a public charging station (15% above the average cost of supply) [25].

While it is beyond the scope of this chapter to examine in detail various challenges and opportunities with regard to charging infrastructure for electric vehicles, a few considerations for policy makers and industry representatives are outlined below:

- Adopt the 'hub' (cities above 4 million) and 'spoke' (national highways including major freight corridors) strategy already acknowledged within FAME 2 to channeling public and private financing for EVs. India could reflect these within its nationally determined contributions (NDCs) under the Paris Agreement and also tap into available sources of multilateral financing for charging infrastructure including agencies such as the Climate financing channels of the World Bank[2] and International Finance Corporation [27], the Asian Development Bank, Global Environment Facility (GEF), United Nations Development Programme as well as other bilateral donors. Innovative carbon credit financing could also be explored for individual charging stations or a cluster of charging stations such as that availed of by the Delhi metro after it commenced operations.
- Among heavy vehicles, in addition to buses, there should also be an emphasis in the medium to long term for deployment of electric freight vehicles including long-range trucks and smaller delivery vehicles. These could complement rail freight traffic and provide last mile connectivity from rail freight hubs to final destinations within a selected radius. Further, they could help address conventional truck-related emissions and pollution with cities as well.
- Ensure, based on a techno-economic assessment and feasibility study, an adequate number of battery swapping facilities along high-density routes (either stand-alone or integrated as part of charging stations). Presently, FAME-2 makes swapping facilities optional for charging stations along highways and does not make it mandatory within cities. Swapping could enable faster turnaround times for the large numbers of vehicles plying Indian highways (going by the current charging times required) as well as greater revenue flows for charging stations. Services such as 'battery on demand'-whereby fully charged batteries could be delivered by swapping stations to stranded vehicles along highways-could be explored further in order to lower range anxiety and boost EV demand. EV manufacturers globally could accommodate swapping flexibility into their battery and vehicle design in order to enable battery leasing and lower upfront vehicle costs [17]. While some studies point to higher costs of operating swap stations in China compared to fast charging [28], with adequate global scale and further innovation such costs may come down. It is likely, however, that for four-wheeler vehicles may be deployed at a bigger scale only after a minimum penetration of EVs has already happened.

---

[2]The World Bank has a dedicated Climate Change Trust Fund for Infrastructure (CCTFI) under the Public–Private Infrastructure Advisory Facility (PPIAF) which it administers and is supported by donors.

- Explore ways of enabling Indian railways to get involved in battery charging and swapping services on site or through leasing of railway land close to highways to private operators. This could introduce a new revenue stream for railways as well, once EV road freight traffic starts increasing.
- Explore ways and means of ensuring that charging stations (including battery swap stations) run as much as possible on renewable energy through on-site or grid-based renewable energy. The techno-economic feasibility of mass-scale charging of EV batteries at suitable sites by renewable energy power producers could also be explored as means of ensuring full utilization of clean energy generated by renewable energy producers in addition to that dispatched into the grid.

# Reference

1. M. Crippa, G. Oreggioni, D. Guizzardi, M. Muntean, E. Schaaf, E. Lo Vullo, E. Solazzo, F. Monforti-Ferrario, J.G.J. Olivier, E. Vignati, Fossil $CO_2$ and GHG emissions of all world countries—2019 Report, EUR 29849 EN, Publications Office of the European Union, Luxembourg, 2019. ISBN 978-92-76-11100-9. https://doi.org/10.2760/687800, JRC11761. https://edgar.jrc.ec.europa.eu/overview.php?v=booklet2019&dst=GHGemi#. Accessed 18 June 2020
2. S. Dhar, P.R. Shukla, M. Pathak, India's INDC for transport and 2 °C stabilization target. Chem. Eng. Trans. **56**, 31–36 (2017). https://doi.org/10.3303/CET1756006
3. S. Singh, T. Mishra, R. Banerjee, *Greenhouse Gas Emissions in India's Transport Sector: Climate Change Signals and Response* (Springer, Berlin, 2019), pp 197–209 https://link.springer.com/chapter/10.1007/978-981-13-0280-0_12. Accessed 18 June 2020
4. A. Sharma, C.E. Chung, Climatic benefits of black carbon emission reduction when India adopts the US on road emission level. Future Cities Environ. **1**, 13 (2015). https://doi.org/10.1186/s40984-015-0013-8. Accessed 18 June 2020
5. S. Anenberg, J. Miller, D. Enzed, R. Minjares, A global snapshot of the air pollution-related health impacts of transportation sector emissions in 2010 and 2015, in *The International Council on Clean Transportation* (2019). https://theicct.org/publications/health-impacts-transport-emissions-2010-2015. Accessed 18 June 2020
6. A. Roychowdhury, G. Dubey, *The Urban Commute: and How it Contributes to Pollution and Energy Consumption* (Centre for Science and Environment, New Delhi, 2018). https://www.cseindia.org/the-urban-commute-8950. Accessed 18 June 2020
7. T.V. Ramachandra, K. Shwetmala, Emissions from India's transport sector: statewise synthesis. Atmos. Environ. **43**, 5510–5517 as cited in Office of International and Tribal Affairs U.S. Environmental Protection Agency, Reducing Black Carbon Emissions in South Asia: Low Cost Opportunities (2009)
8. Measures to Control Air Pollution in Urban Centres of India: Policy and Institutional framework. The Energy and Resources Institute (TERI). Policy Brief. February 2018. https://www.teriin.org/sites/default/files/2018-03/policy-brief-air-pollution-in-urban-centres-of-India.pdf. Accessed 18 June 2020
9. United Nations, Department of Economic and Social Affairs, Population Division. World Urbanization Prospects: The 2014 Revision, Highlights (2014). https://population.un.org/wup/Publications/Files/WUP2014-Highlights.pdf. Accessed 18 June 2020
10. A. Raman, S. Shukla, Vehicle Inspection programme needs an overhaul. Down to Earth (2018). https://www.downtoearth.org.in/news/air/vehicle-inspection-programme-needs-an-overhaul-62052. Accessed 18 June 2020

11. Z. Shao, *Bharat Stage VI Emission Standards: Mission Not Impossible*. The International Council on Clean Transportation (ICCT), 2020. https://theicct.org/blog/staff/bharat-stage-vi-mission-not-impossible. Accessed 18 June 2020
12. A. Roychowdhury, *Bharat Stage VI: India leapfrogs today and it is no Fool's day*. Down to Earth Blog (2020). https://www.downtoearth.org.in/blog/air/bharat-stage-vi-india-leapfrogs-today-and-it-is-no-fool-s-day-70155. Accessed 18 June 2020
13. T. Dallman, *Leapfrogging an outdated standard puts India on par with global leaders in control of vehicle emissions*. Staff Blog. The International Council on Clean Transportation (ICCT), 2016. https://theicct.org/blogs/staff/india-leapfrogging-an-outdated-standard-to-bharat-stage-VI. Accessed 18 June 2020
14. NITI Aayog and Rocky Mountain Institute, *Goods on the Move: Efficiency & Sustainability in Indian Logistics* (2018). https://niti.gov.in/writereaddata/files/document_publication/Freight_report.pdf. Accessed 18 June 2020
15. Indian Railways' dedicated freight corridor: 6 things to know. The Economic Times. 27 February 2020. https://economictimes.indiatimes.com/industry/transportation/railways/indian-railways-dedicated-freight-corridor-6-things-to-know/articleshow/74337277.cms. Accessed 18 June 2020
16. International Energy Agency, *The Future of Rail: Opportunities for Energy and the Environment* (2019). https://www.iea.org/reports/the-future-of-rail. Accessed 18 June 2020
17. Federation of Indian Chambers of Commerce and Industry (FICCI) and Rocky Mountain Institute, *Enabling the Transition to Electric Mobility in India* (2017). https://rmi.org/wp-content/uploads/2017/11/report_electric_mobility_india_FICCI_RMI.pdf. Accessed 18 June 2020
18. Society for Manufacturers of Electric Vehicles (SMEV). https://www.smev.in/ev-industry as cited in Jain, A. (2019). The Clean Makeover of India's Transportation Systems: Reasons for Optimism. Blog Article. The Climate Group. https://www.theclimategroup.org/news/clean-makeover-india-s-transportation-systems-reasons-optimism. Accessed 18 June 2020
19. NITI Aayog & World Energy Council, *Zero Emission Vehicles (ZEVs): Towards a Policy Framework* (2018). https://niti.gov.in/writereaddata/files/document_publication/EV_report.pdf. Accessed 18 June 2020
20. India's Rickshaw Revolution Leaves China in the Dust, Bloomberg News. 25 Oct 2018. https://www.bloomberg.com/news/features/2018-10-25/india-s-rickshaws-outnumber-china-s-electric-vehicles. Accessed 18 June 2020
21. Several cities block electric rickshaw registrations. The Economic Times. 24 Mar 2020. https://economictimes.indiatimes.com/industry/auto/auto-news/several-cities-block-electric-rickshaw-registrations/articleshow/74793376.cms?from=mdr. Accessed 18 June 2020
22. World Economic Forum and Ola Mobility Institute, *EV-Ready India Part 1: Value-Chain Analysis of State EV Policies* (2019). https://www3.weforum.org/docs/WEF_EV_Ready_India.pdf. Accessed 18 June 2020
23. J. Callaghan, S. Rodakiya, *Boosting EV Growth in India: Missing Pieces in the Policy Puzzle*. Staff Blog. International Council on Clean Transportation (ICCT), 2019. https://theicct.org/blog/staff/boosting-ev-growth-india-20191028. Accessed 18 June 2020
24. Bureau of Energy Efficiency (BEE) and Ernst and Young LLP, *Propelling Electric Vehicles in India: Technical Study of Electric Vehicles and Charging Infrastructure* (2019). https://beeindia.gov.in/. Accessed 18 June 2020
25. Government of India, *Charging Infrastructure for Electric Vehicles—Guidelines and Standards* (2018). https://powermin.nic.in/sites/default/files/webform/notices/scan0016%20%281%29.pdf. Accessed 18 June 2020
26. Government of India, *Amendments in Model Building By-Laws (MBBL-2016) for Electric Vehicle Charging Infrastructure* (2019). https://pibphoto.nic.in/documents/rlink/2019/feb/p201921501.pdf. Accessed 18 June 2020
27. See for example International Finance Corporation, *Climate Investment Opportunities in Emerging Markets: An IFC Analysis* (2016). https://www.ifc.org. Accessed 18 June 2020

28. J.M. Grütter, K.J. Kim., E-mobility options for ADB developing member countries (2019) Asian Development Bank Working Paper Series No. 60. https://www.adb.org/sites/default/files/publication/494566/sdwp-060-e-mobility-options-adb-dmcs.pdf. Accessed 18 June 2020

# Opportunities for Future Tailpipe Emissions Regulation of Light-Duty Vehicles Within the European Union

Anna Krajinska

**Abstract** The introduction of on-road real driving emissions (RDEs) tests for type approval and in-service emissions compliance has driven the reduction of on-road tailpipe emissions from light-duty vehicles within the European Union. However, further reductions of tailpipe emissions are possible. With the European Commission currently deliberating whether, after Euro 6, a further emission standard is necessary, this chapter discusses potential areas of improved emissions regulation in relation to light-duty vehicles. A general overview is given of the drivers for further reductions of tailpipe emission limits, the regulation of currently unregulated pollutants, improvements to the RDE test procedure as well as increases to durability and in-service conformity requirements.

## 1 Introduction

Direct tailpipe emission limits for light- and heavy-duty vehicles have been in place across the Member States of the European Union (EU) since the implementation of the first 'Euro' vehicle emission standards in 1992. Since their implementation, emissions of key pollutants from road transport in the EU have been in decline. Air pollution data collected by the European Environmental Agency between the years 2000 and 2017 indicates particularly large reduction has occurred for emissions of carbon monoxide (CO) and non-methane volatile organic compounds (NMVOC) representing a decrease of almost 70% over 17 years. Reductions in emissions of black carbon, particulate matter (both $PM_{2.5}$ and $PM_{10}$) as well as nitrogen oxides (NOx), were slightly more modest of between 40 and 60%. This has occurred despite road passenger and freight volumes gradually increasing during this period [1].

However, despite these reductions and almost three decades of vehicle emissions regulation, including six progressively more stringent 'Euro' standards, air pollution from road transport continues to negatively impact air quality across the EU. In

A. Krajinska (✉)
London, UK
e-mail: krajinska.anna@gmail.com

2017, the road transport sector was the biggest source of NOx emissions in the EU accounting for 39% of total emissions, as well as the second biggest source of black carbon, carbon monoxide and primary $PM_{2.5}$ accounting for 28%, 19% and 11%, respectively. The problem is particularly acute in cities and urban areas where road traffic volumes are high, with numerous cities across the EU infringing the ambient air quality limit for nitrogen dioxide ($NO_2$), of which road transport is the biggest source, set out in the EU Ambient Air Quality Directive (EAAQD) [1]. As of October 2019, the European Commission had initiated infringement procedures against fourteen out of twenty-eight EU Member States for persistent exceedances of the EAAQD $NO_2$ limit [2].

During the first and second quarter of 2020, restriction on movement imposed by Member States across the EU, aimed at limiting the spread of Covid-19, resulted in temporarily large reductions in road traffic volumes and subsequently large reductions in ambient air concentrations of $NO_2$. Upon introduction of severe restrictions on movement beginning on the 24th of March 2020, $NO_2$ concentrations in London decreased, compared to the first ten weeks of the year by on average 21.5% across London's roads and 14% at urban background sites, with reductions of up to 55% recorded at busy roadsides [3]. Similarly, average $NO_2$ concentrations in Barcelona, Madrid and Lisbon decreased by 40–56% week on week upon introduction of lockdown measures [4] suggesting that reductions in traffic volumes or stricter emissions regulation targeted at reducing the quantity of pollutants, especially $NO_2$, emitted from road vehicles could be effective in significantly improving air quality in Europe's cities.

This chapter discusses the opportunities for strengthening the regulation of tailpipe emissions for light-duty (LD) vehicles as part of the development of post-Euro 6 emission standards in particular discussing lower and fuel neutral emission limits, unregulated pollutants, changes to the real driving emissions (RDEs) test procedure as well as increased durability and in-service conformity requirements.

## 2 Opportunities for Future Tailpipe Emissions Regulation of Light-Duty Vehicles Within the European Union

The introduction of on-road Real Driving Emissions (RDE) testing, not-to-exceed on-road emissions limits for NOx and particle number (PN) as well as in-service conformity testing requirements for Member States as part of Euro 6d-temp and 6d regulation appears to have reduced the emissions of NOx and PN from most LD vehicles to within the applicable legal limits, when tested inside the boundaries of the new RDE test procedure [5, 6]. However, the European Commission, who is responsible for vehicle emissions regulation within the EU single market, has acknowledged that after the implementation of Euro 6, more needs to be done to reduce pollution from internal combustion engines as the 'current standards do not sufficiently contribute to the decrease in air pollutant emissions emerging from road transport, required

for the move towards zero-pollution in Europe and the protection of human health'
[7]. The EU's zero-pollution ambition was first outlined in December 2019 as part
of the European Green Deal [8]—the EU's roadmap for making the EU's economy
sustainable, including for transport. However, it is not yet clear in what manner zero-
pollution will be applied to transport, particularly in relation to any future vehicle
emissions regulation.

At the time of writing, the European Commission was in the process of conducting
an impact assessment to determine if another emission standard is necessary. The
Commission has also put together a consortium of experts under the title of CLOVE,
who are in the process of critically reviewing the effectiveness of the Euro 6 regulation and who will provide recommendations on how the regulation could be further
improved. If the Commission decides to move forward with a new emission standard, it is expected that a draft will be ready in the fourth quarter of 2021, at the
earliest [9], with legal implementation not expected prior to 2025. The Commission
has also signalled that the next EU vehicle emissions regulation will likely be its last
for conventional combustion engines prior to the transition to low and zero-emission
options, potentially indicating large changes in order to future-proof the regulation
and to ensure that it can accelerate the shift to sustainable and smart mobility, a key
goal of the European Green Deal [10]. The most impactful areas of focus, for the
new regulation, in relation to the control of tailpipe emissions are likely to include:
the magnitude of the emissions limits, new pollutants for regulation, improvements
to the RDE test procedure as well as increased durability and in-service conformity
testing requirements.

## 2.1   Lower and Fuel Neutral Emission Limits

Of priority interest for post-Euro 6 regulation is the reduction of emissions limits of
currently regulated pollutants, given that the last Euro 6 limits [(EC) 715/2007] were
finalised in 2007 and significant engine and after treatment technological progress
has occurred since. As well as a reduction in limits, the harmonisation of limits for all
power trains and fuel types is also on the table, including between LD (Euro 6) and
heavy-duty (HD) (Euro VI) vehicles which are subject to different emissions limits
at present. This is driven by the Commission's key policy objective of complexity
reduction for post-Euro 6 regulation [7]. Euro 6 also discriminates between different
fuels and power trains, in contrast to the fuel neutral approach adopted in the USA
and in China.[1] Within the LD category, passenger cars and light commercial vehicles
(LCVs) are subject to different limits, with three different sets of limits for LCVs
depending on gross vehicle mass.

In terms of regulation of specific pollutants, only NOx, carbon monoxide (CO),
particulate matter (PM) and hydrocarbons (HC) are regulated for all LD vehicles
including port-fuel and direct-injection petrol, diesel and natural gas. Even for these

---

[1] Beginning from China 6a, nationwide implementation of which begins from the 1st July 2020.

pollutants, different limits apply for positive and compression ignition vehicles; the
NOx limit is 25% higher for compression ignition cars and the CO limit for compression ignition is half that for positive ignition cars. Euro 6 emission limits for LD
vehicles are presented in Table 1. Since 2017 vehicle emissions at type approval
are tested and regulated on the World Harmonised Light-duty Test Cycle (WHTC),[2]
replacing the old New European Drive Cycle (NEDC) with a longer, faster and more
dynamic test which is more representative of real-world driving in the EU [11] limits
for all regulated pollutants apply on the WHTC test. Beginning with the introduction of the on-road RDE test from September 2017[3] at Euro 6d-temp, not-to-exceed
emission limits for NOx and PN apply on the RDE tests (CO is measured but not
regulated). RDE not-to-exceed limits are calculated by applying a conformity factor
(CF) to the Euro 6 emission limits.[4] The CF for NOx at 6d-temp is 2.1, reduced to
1.43 at 6d (beginning in January 2020). For PN, the CF is 1.5, and these are further
discussed in Sect. 2.3.

Use of new technologies for emission control such as passive NOx absorbers for
cold-start NOx control, dual dosing urea selective catalytic reduction (SCR) systems
allowing for multiple independently dosed SCR bricks strategically located in the
exhaust and use of advanced SCR coatings allows for optimal NOx conversion efficiency under a wide range of driving conditions and allows diesel cars to achieve
NOx emissions comparable to petrol cars. A Euro 6b diesel car retrofitted and recal-
ibrated by the Association for Emissions Control Catalysts including the fitting of
a dual dosed SCR system and mild 48 V hybridisation achieved NOx emissions of
~40 mg/km on both laboratory and RDE tests, half of the 80 mg/km diesel NOx
limit [12]. Analysis by Transport & Environment of publicly available type approval
RDE emissions of 6d-temp diesel cars in 2019 revealed that at least 59 emitted less
than 20 mg/km during both urban driving and the entire RDE test [13] indicated that
some diesel cars on sale today are already capable of emitting less than a quarter of
current NOx emission limit. This combined with the decrease in passenger car NOx
emissions limits in China to 35 mg/km in 2023 (also regulated on the WHTC and
RDE tests), and the progressive decrease in the fleet average limit in California to
19 mg/km for 2025 indicates that a substantial decrease to EU NOx emission limits
should be feasible.

Similarly, a review of the PN emission limit of $6 \times 10^{11}$/km, applicable to diesel
and direct-injection petrol vehicles may be advisable. The current limit is based on
tests undertaken more than 13 years ago, by the UN Particle Measurement Programme
working group [14], on diesel particle filter equipped diesel vehicles. Despite average
PN emissions of $2 \times 10^{11}$/km from the tested vehicles, the PN emission limit was
set at $6 \times 10^{11}$/km, the highest result measured from a vehicle which was fitted with
a high porosity cordierite particle filter, not used on any of the other tested vehicles
and rarely use on LD diesel vehicles today with the higher filtration efficiency silicon
carbide preferred. Newer Euro 6 diesel vehicles are capable of significantly lower

[2]Beginning from September 2017 for new type vehicles and September 2018 for all vehicles.
[3]Start date for use of the RDE test for type approval depends on the light-duty vehicle category.
[4]Start date for use of the RDE test for ype-approval depends on the light-duty vehicle category.

PN emission than $6 \times 10^{11}$/km, with several studies reporting Euro 6 diesel vehicles which were capable of emissions of less than $1 \times 10^{10}$/km during WLTC [15] and RDE [16, 17] based tests. While tailpipe regulation of black carbon emissions is not expected to be introduced in the future, as a decrease in the PN limit is expected to reduce black carbon emissions also.

The PN limit does not apply to LD natural gas or port-fuel injection (PFI) petrol vehicles. The reasoning behind this exclusion pertained to most Euro 5 port-fuel injection petrol [18] and natural gas engines being capable of meeting the PN limit, and therefore, a limit was deemed unnecessary. However, on-road testing [19] indicates that Euro 6 PFI vehicles can exceed the $6 \times 10^{11}$/km PN limit. When tested on routes compliant with RDE legislation, two of the cars emitted on or above the PN limit with PN emission of $6 \times 10^{11}$ and $2.1 \times 10^{12}$/km. Outside of RDE boundary conditions, on a more dynamic test, PN emissions increased dramatically with one vehicle emitting $1 \times 10^{13}$/km [19]. The same study also tested a Euro 6 compressed natural gas LCV which under some driving conditions emitted up to $1 \times 10^{12}$/km. High PN emissions from these vehicles suggest that future extension of PN limits to include PFI and CNG vehicles may be desirable. There is a further risk that if a decrease in the PN limit is not accompanied by an extension of the limit to CNG and PFI vehicles, manufacturers may shift production to these vehicles rather than to engineer diesel or petrol vehicles capable of meeting the future lower PN limit which could result in an increase, rather than a decrease in EU's LD fleet's PN emissions.

## 2.2  Tailpipe Limits for Unregulated Pollutants

The EU is rather conservative in the range of pollutants regulated through a direct tailpipe emission limit for LD vehicles with only nitrogen oxides (NOx), particulate matter (PM), particle number (PN) and hydrocarbons (HC, either as total hydrocarbons and non-methane hydrocarbons for positive ignition engines or through a combined NOx and hydrocarbon limit for compression ignition engines) regulated. However, as shown in Table 1, these limits do not apply uniformly to all LD vehicles. Conversely, the US regulation includes direct tailpipe limits on emissions of formaldehyde as well as the greenhouse gases nitrous oxide ($N_2O$) and methane ($CH_4$). China also regulates $N_2O$.

The CLOVE consortium has suggested five new pollutants, which can be effectively limited through a tailpipe limit, for inclusion into the next Euro emission standard: smaller than 23 nm solid particles, ammonia, formaldehyde, nitrous oxide and methane along with potential revisions to the regulation of $NO_2$ (currently regulated through a combined NOx limit) and the total hydrocarbon limit to non-methane organic gases (NMOG)—the approach used in US regulation. The CLOVE consortium is focused on the measurement and regulation of these pollutants during RDE tests [20] which would dramatically expand the amount of pollutants measured and regulated during on-road driving.

**Table 1** Euro 6 emission limits for light-duty vehicles[a]

| Pollutant | Passenger cars, light commercial vehicles (≤1305 kg) | | Light commercial vehicles (1305–1760 kg) | | Light commercial vehicles (1760–3500 kg) | |
|---|---|---|---|---|---|---|
| | Positive ignition | Compression ignition | Positive ignition | Compression ignition | Positive ignition | Compression ignition |
| NOx (mg/km) | 60 | 80 | 75 | 105 | 82 | 125 |
| CO (mg/km) | 1000 | 500 | 1810 | 630 | 2270 | 740 |
| NMHC (mg/km) | 68 | N/A | 90 | N/A | 108 | N/A |
| THC (mg/km) | 100 | N/A | 130 | N/A | 160 | N/A |
| HC + NOx (mg/km) | N/A | 170 | N/A | 195 | N/A | 215 |
| PN (#/km) | $6 \times 10^{11}$ [b] | $6 \times 10^{11}$ | $6 \times 10^{11}$ [b] | $6 \times 10^{11}$ | $6 \times 10^{11}$ [b] | $6 \times 10^{11}$ |
| PM (mg/km) | 4.5 | 4.5 | 4.5 | 4.5 | 4.5 | 4.5 |

[a]Regulation (EC) No. 715/2007
[b]Applies only to direct-injection vehicles

The PN emissions limit for solid, >23 nm ($PN_{23}$), particles have been effective in driving the adoption of particle filters for diesel and petrol LD vehicles, especially once limits were enforced on RDE tests. However, recent research suggests that a large number of solid <23 nm ($PN_{10-23}$) particles are also emitted, in many cases exceeding the emissions of $PN_{23}$. The number of $PN_{10-23}$ emitted is highly dependent on the fuel and engine technology used in the vehicle. Johnsson et al. [21] found that for the majority of LD petrol and diesel vehicles, the ratio of $PN_{23}$ to $PN_{10-23}$ was 1:1–4; however, for CNG vehicles, the ratio can be close to 1:30. In line with these results, Giechaskiel et al. [22] observed $PN_{10-23}$ constituting 60–100% of total PN emissions for natural gas and PFI engines and <50% for diesel engines (with or without a DPF). Krajinska et al. [16] reported an increase of between 11% and 184% for two 6d-temp diesels during RDE-based laboratory cycles.

The extension of particle measurement to include $PN_{10-23}$ has been investigated by three European Commission funded projects: DownToTen, Sureal-23 and PEMS4Nano as well at the UN Particle Measurement Programme working group. A UN Global Technical Regulation (GTR) for measuring particles >10 nm in the laboratory is currently in the process of being finalised, and the submission of a proposal for PEMS-based measurement of >10 nm particles is expected June 2021 [23]. It is widely expected that this will be included in any post-Euro 6 legislative proposal, given a GTR regulation should be finalised by the time a draft of the post-Euro 6 regulation is ready towards the end of 2021 and as a placeholder for the measurement accuracy of 10 nm particles has been included in the HD Euro VI Step E regulation, (EU) 2019/1939.

The risk of high ammonia ($NH_3$) tailpipe emissions from LD vehicles is increasing with widespread use of SCR for diesel LD vehicles and increasing loadings of platinum group metals in three-way catalysts [24] required for more stringent NOx control. While $NH_3$ tailpipe emissions are regulated for HD vehicles and ammonia storage catalysts are widely used to reduce $NH_3$ emissions, this is not the case for LD vehicles. While agriculture is the dominant source of ammonia emissions in the EU, in cities road vehicles can be the dominant source [25]. The main driver for regulating ammonia emissions is its potential to form secondary particles, contributing to $PM_{2.5}$ [26]. Ammonia emissions in excess of 30 mg/km have been measured for a diesel vehicle on RDE-based laboratory cycles when undergoing DPF regeneration [16], and Suarez-Bertoa et al. [27] measured emissions of 21–48 mg/km during RDE testing of Euro 6 petrol vehicles as well as emissions of up to 66 mg/km for one Euro 6 CNG-LCV tested. CLOVE has assessed the on-road measurement of ammonia emissions as already possible using a laser diode gas analyser (LDS) and the use of Fourier transform infrared spectroscopy (FTIR) or a quantum cascade laser (QCL) for this purpose as feasible.

Nitrous oxide and methane are potent greenhouse gases with global warming potential of 265 and 28 times that of $CO_2$, respectively [28]. Methane also oxidises to form ground level ozone [29], a greenhouse gas and pollutant associated with negative respiratory health effects [30]. Gas vehicles are at particular risk of high methane emissions, especially at cold-start when exhaust gas temperatures are low, due to the high three-way catalyst temperatures required for efficient methane oxidation [31]. Methane is already directly regulated during on-road testing for HD vehicles and measured using gas chromatography with flame-ionising detection (GC-FID) and non-methane cutter flame ionising detector (NMC-FID); however, this approach is not considered suitable for LD due to the requirement for gas bottles inside the vehicle. FTIR is promising for both methane and $N_2O$.

Formaldehyde emissions increase with an increasing percentage of ethanol in the fuel [32]. While current EU sales of E85 vehicles are low, as the EU makes progress towards its goal of net zero $CO_2$ emissions by 2050 [8], potentially soon to be enshrined in law [33], the use of bioethanol to power the EU LD fleet may increase and subsequently with this so may formaldehyde emissions. This is concerning as formaldehyde is classified as carcinogenic [34]. The majority of petrol sold in the EU at present is blended with 5% ethanol (E5); however, 10% ethanol blends made up 9% of the EU petrol market in 2016 [35]. The USA, Brazil and Korea—markets where the use of higher ethanol blends is more common—already set tailpipe emission limits for formaldehyde; however, these are exclusively regulated on laboratory test cycles. As the measurement of formaldehyde on the road using FTIR is promising, the EU may become the first region to regulate formaldehyde emissions during on-road driving.

It is not yet clear if CLOVE will advocate for the reduction of the combined nitrogen oxide (NO) and nitrogen dioxide ($NO_2$), known as the NOx limit, in order to drive the reduction of $NO_2$ emissions or suggest a separate limit for $NO_2$. There are arguments for both approaches; $NO_2$ is the fraction of NOx harmful to human health; however, (NO) is also oxidised to $NO_2$ [36], albeit not fully, due other competing reactions, suggesting that emissions of both should be decreased in order to further reduce ambient concentrations of $NO_2$. However, large differences in the percentage of $NO_2$ emitted from Euro 6 diesel cars, remote sensing measurements [37] put this at between 4% and 41%, could justify introducing a separate limit for $NO_2$ alongside a combined NOx limit. The eventual choice in how to best regulate NOx emissions will likely come down to the expected benefit to ambient air quality and human health versus the cost and the increased regulatory complexity of two separate limits. Similarly, a shift away from regulating total hydrocarbons (THC) to NMOG (the approach in the USA) would result in other organic gases such as aldehydes, not just hydrocarbons, to be regulated. The present EU procedure for calculating THC does account for organic gases, but in some cases, it may underestimate the total amount particularly for high ethanol blends [38], and therefore, the US procedure for calculating NMOG emissions may be preferable.

## 2.3   Expansion of the RDE Test

The guiding principles and objectives of the CLOVE consortium for post-Euro 6 regulation focus on guaranteeing that vehicles are as clean as possible under all driving conditions, over their entire useful life, to be ensured through RDE testing and on-board monitoring [20]. While the present RDE test procedure covers a much wider range of engine map operating conditions than the laboratory-based NEDC [39] or WLTP tests [40], vehicles are only effectively required to meet the emission limits for NOx and PN within the boundaries of the RDE test procedure. Not-to-exceed (NTE) emission values on RDE tests for NOx and PN are calculated by multiplication of the Euro 6 emission limit by the conformity factor (CF), no limit exists for CO despite measurement of this pollutant. At Euro 6d-temp, beginning in 2017, a CF of 2.1 applies to NOx resulting in an NTE of 168 mg/km for diesel and 126 mg/km for petrol passenger cars. The CF is reduced to 1.43 at 6d (114 mg/km for diesel, 86 mg for petrol). A CF of 1.5 for PN applies to 6d and 6d-temp resulting in an NTE of $9 \times 10^{11}$/km. The European Commission is legally obliged to review the CFs on an annual basis, based on improvements to PEMS technology. The latest proposal by the Joint Research Centre suggests reducing the CF for NOx to 1.32 [41], to date no reduction in the PN CF, has been proposed.

The RDE test procedure was designed to cover a wide range of normal driving conditions on EU roads including cold-start and is suitable for testing of all passenger and commercial LD vehicles including plug-in hybrids. In comparison with the previous laboratory-based NEDC or the current WLTP test, there is no fixed test route. There are only 'boundary conditions' which have to be met during the test: the test must last between 90 and 120 min and is split into urban, rural and motorway phases, with a minimum driving distance of 16 km in each, further details are provided in Table 2. Maximum altitude (700 m), maximum cumulative altitude gain

**Table 2** Urban, rural and motorway phase requirements for a valid RDE test[a]

| Valid RDE trip requirements | | | |
|---|---|---|---|
| | Urban | Rural | Motorway |
| Distance of total test (%) | 29–44 | 23–43 | 23–43 |
| Speed (km/h) | ≤60 | 60–90 | >90 |
| Average speed (km/h) | 15–40 | N/A | N/A |
| Total stop time (%) | 6–30% | N/A | N/A |
| Individual stop time (s) | ≤300 | N/A | N/A |
| Speed > 100 km/h (min) | N/A | N/A | 5 |
| Speed > 145 km/h (%) | N/A | N/A | 3% |

[a]EU 2016/427, (EU) 2016/646

(1200 m/100 km), minimum and maximum driving dynamics and temperature boundaries (0–35 °C, with a derogation to 3–30 °C until January 2020) apply. Temperature and altitude boundaries can be extended to between −7 and 35 °C (−2 to 35 °C until January 2020) and up to 1300 m; however, on these tests, applicable emission limits are multiplied by 1.6. The last RDE legislative package (EU) 2018/1832 also introduced mandatory independent in-service conformity testing requirements beginning from January 2020, and these are further discussed in Sect. 2.4.

Falling outside of RDE test boundaries are after treatment regeneration events including diesel particulate filter (DPF) regenerations (unless a regeneration occurs on two consecutive tests), which involve the burn-off of soot accumulated inside the filter and have been reported to occur as often as every 200 km [42]. Laboratory testing of a pre-RDE Euro 6 car during regeneration [43] indicated large increases in emissions of both regulated (CO, PM, HC) and unregulated pollutants (NH$_3$, SO$_2$) with PN and NOx exceeding the applicable emission limits. This corroborates with average PN emissions from DPF fitted diesel vehicles reported by Andersson et al. [21] of $9 \times 10^{11}$/km and DPF regeneration data collected by Giechaskiel et al. [22] of Euro 6 vehicles, 3 out of 6 exceed the PN limit with emission of $1.9$–$6.9 \times 10^{12}$/km. Testing by Krajinska et al. [16] indicates that high PN emissions during DPF regeneration are a continuing problem for RDE approved vehicles with regenerating emissions of 2 Euro 6d-temp diesel cars on RDE-based laboratory cycles of $7.9 \times 10^{11}$–$1.3 \times 10^{12}$/km. Further analysis suggests that excluding PN emissions from regenerating tests ignores 60–99% of the total particles emitted by the two cars. While excess NOx emissions on these test did not occur, NOx emissions in excess of double the limit were reported during WLTC testing at 14 °C of a Euro 6d-temp diesel car [42].

The CLOVE consortium suggests that DPF regenerations may be regulated as part of special operation RDE (SORDE). These tests would be conducted on the road and could be expended to cover any driving condition which is not covered by the present RDE test including after treatment regeneration, short distance driving, harsh accelerations, high-speed driving as well as high temperature conditions [44]. Introduction of SORDE testing could result in a significant increase in the amount of driving conditions tested and covered by emissions limits similar to the expansion of current RDE test boundaries.

At present maximum and minimum driving dynamics requirements for RDE tests are limited through the 95th percentile of positive products of vehicle velocity and acceleration [$v * a_{pos}$ (95%)] and relative positive acceleration (RPA), respectively. Both exist to prevent overly aggressive or gentle driving from resulting in too high or low an emission result, which would be considered unrepresentative of vehicle use in the EU. However, an analysis of the driving of Groupe PSA customers in Europe suggests that the $v * a_{pos}$ (95%) boundary may unduly classify normal driving as 'too aggressive'; during rural and motorway driving, most customers were found to be close to or above the $v * a_{pos}$ (95%) boundary, especially those driving high powered or sport utility-vehicles [45]. Furthermore, on-road tests show light-duty vehicles can exceed the emission limit when driven more dynamically than $v * a_{pos}$ (95%) allows even for RDE certified vehicles, all 3 diesel 6d-temp cars tested by Suarez-Bertoa et al. [19] exceeded the NOx limit, with emissions of up to 338 mg/km; similarly, Giechaskiel et al. [46] reported emissions of up to 200 mg/km for a Euro 6d-temp diesel car.

Emissions in excess of the PN limit were also measured under dynamic driving conditions for a 6d-temp petrol car [47] suggesting that both petrol and diesel vehicles are at risk of emissions non-compliance when driven more dynamically than the regulation allows. Limits on driving dynamics may have been a practical consideration when RDE testing was considered for use at type approval only; limits on minimum/maximum driving dynamics would ensure that the type approval test was representative of the average use of a car in the EU. However, with the introduction of in-service conformity testing requirements with the last step of the RDE regulation (EU) 2018/1832, which includes periodic independent in-use RDE testing requirements, minimum/maximum driving dynamics requirements for RDE tests effectively restrict the conditions under which LD vehicles can be tested as part of in-service conformity testing and therefore effectively limits the conditions under which emission limits must be met. As such, a review of driving dynamics in future-regulation may be necessary.

Limits on maximum cumulative gain (1200 m/100 km) and a maximum limit on altitude difference between the start and end of the RDE test (±100 m) risk emission limits being exceeded when driving in hilly or mountainous regions. In some EU countries, it can be difficult to drive a valid trip as even in moderately hilly regions the cumulative altitude gain can be considerable [48], an RDE route developed by PSA around Paris, a region not known for its steep inclines almost exceeded the cumulative altitude limit, reaching 1100 m/100 km [49] indicating that these requirements are likely to be too restrictive. Similarly, while the present temperature boundaries (especially extended boundaries) cover a temperature range which is experienced across Europe throughout the year, it does not cover all weather conditions especially those experienced in winter in Northern Europe and summer in Southern Europe where temperatures can be far in excess of even the extended boundary RDE temperatures.

There is flexibility in extending the RDE boundaries given that some PEMS equipment is capable of measuring emissions within a temperature range of between −10 and 45 °C and an altitude of 3000 m.

Currently, raw emissions results from RDE tests are post-processed using the moving average window (MAW) method. However, the final emission result is quite sensitive to the results of the WLTP test used for the calculation. The European Commission's Joint Research Centre has taken the approach of using raw emissions results in several of their research papers suggesting that there is potential to move towards raw emission results in the future.

## 2.4 In-service Conformity and Emissions Durability Requirements

The current emission durability requirements require emission limits to be met by a vehicle for the first 160,000 km, as verified at type-approval; however, prior to the introduction of mandatory RDE in-service emission testing requirements to be undertaken by the Commissions and Member States and made public on an annual basis, with the last step of the RDE regulation (EU) 2018/1832, only manufacturers were required to periodically test the emissions performance of in-use vehicles and only on the laboratory test cycle. While this was overseen by the Member State type approval authorities, the use of the NEDC test cycle, lack of on-road testing and lack of public scrutiny of the emission results were some of the key factors that were attributed to the dieselgate scandal, in which many EU car manufacturers were implicated in engaging in widespread emission cheating.

The introduction of mandatory RDE in-service conformity (ISC) testing, starting from September 2019, was introduced as a safeguard to help ensure that vehicles actually meet the emission limits on the road throughout their lifetime. However, the maximum LD vehicle age and mileage restrictions for ISC tests, of between 15,000 and 100,000 km and an age of five years or younger, whichever comes first, fall short of the type-approval 160,000 km durability requirement. Additionally, both the type-approval and in-service durability requirements fall far short of the durability requirements in the USA (240,000 km) and China (200,000 km).[5] Furthermore, the average age of a passenger car in the EU (11.1 years) is more than double the in-service conformity testing requirement, with some Member States reporting an average age of up to 17.3 years [50]. This combined with the Commission's objective of 'keeping air pollutant emissions under control throughout their entire lifetime and in all conditions of use' [7] suggests that in-service conformity and durability requirements are likely to be reviewed as part of any future post-Euro 6 emission standard. CLOVE have also suggested that future in-service conformity testing may encompass a significantly wider range of driving conditions through the inclusion of SORDE testing outside of present RDE testing boundaries [44].

---

[5]From China 6b beginning in 2023.

# 3   Conclusion

The European Union has come far in improving the on-road emissions performance of light-duty vehicles in recent years, particularly in relation to reduction of nitrogen oxide (NOx) emissions. However, further opportunities to reduce tailpipe emissions from internal combustion engines still exist. Particular areas of interest for tightening of future emissions regulation are the application of emission limits uniformly, regardless of fuel or power train technology, the reduction of emission limits for currently regulated pollutants including NOx, particulate matter (PM), particulate number (PN), hydrocarbons and carbon monoxide (CO), and the regulation of currently unregulated pollutants especially-namely smaller than 23 nm particles, ammonia ($NH_3$), nitrous oxide ($N_2O$), methane ($CH_4$) and formaldehyde. Changes to emission limits for the regulation of nitrogen dioxide ($NO_2$) and total hydrocarbons may also be beneficial. Alongside this, there could be improvements to the real driving emissions (RDEs) testing procedure to encompass a wider range of driving conditions as well as extensions of emissions durability and in-service conformity requirements. Improvements to the EU emissions regulation in these areas have the potential to significantly further reduce tailpipe emissions from light-duty vehicles in the EU. The European Commission is currently in the process of determining if another 'Euro' emission standard is necessary, and if so, what improvements should occur. However, due to announcements made by the Commission as part of the European Green Deal, it is widely expected that the process will go ahead with a new legislative proposal expected at the earliest in late 2021.

# References

1. A.G. Ortiz, C. Guerreiro, J. Soares, *Air Quality in Europe-2019 Report* (European Environmental Agency, 2019), https://www.eea.europa.eu/publications/air-quality-in-europe-2019. Accessed 4 Apr 2020
2. European Commission, *Fitness Check of the Ambient Air Quality Directives.* Commission staff working document (2019), https://ec.europa.eu/environment/air/pdf/SWD_2019_427_F1_AAQ%20Fitness%20Check.pdf. Accessed 14 May 2020
3. Environmental Research Group, *The Effect of Covid-19 Lockdown Measures on Air Quality in London in 2020.* Report (King's College London, 2020), https://assets.ctfassets.net/9qe818412nz4/2TM8WJUt2w1cHecdjkVIRQ/2e5a91667d676b3c63f1e748156b68c4/ERG_response_to_Defra.pdf. Accessed 14 May 2020
4. European Environmental Agency, *Air Pollution Goes Down as Europe Takes Hard Measures to Combat Coronavirus* (2019), https://www.eea.europa.eu/highlights/air-pollution-goes-down-as. Accessed 14 May 2020
5. Euro 6 Real Driving Emissions Data, European Automobile Manufacturer's Association. https://www.acea.be/publications/article/access-to-euro-6-rde-monitoring-data. Accessed 14 Apr 2020
6. Euro 6 Real Driving Emissions Monitoring Data (Japan Automobile Manufacturer's Association, INC.), https://www.jama-english.jp/europe/publications/rde.html. Accessed 20 Apr 2020

7. Development of post-Euro 6/VI emission standards for cars, vans lorries and buses. Combined Evaluation Roadmap/Inception Impact Assessment. European Commission, https://ec.europa.eu/info/law/better-regulation/have-your-say/initiatives/12313-Development-of-Euro-7-emission-standards-for-cars-vans-lorries-and-buses. Accessed 27 May 2020

8. The European Commission, *The European Green Deal*. Communication from the Commission (2019), https://eur-lex.europa.eu/legal-content/EN/TXT/?qid=1588580774040&uri=CELEX:52019DC0640. Accessed 15 May 2020

9. European Commission, *European Vehicle Emissions Standards-Euro 7 for Cars, Vans, Lorries and Buses* (2020), https://ec.europa.eu/info/law/better-regulation/have-your-say/initiatives/12313-Development-of-Euro-7-emission-standards-for-cars-vans-lorries-and-buses. Accessed 15 May 2020

10. European Parliament, *A European Green Deal: Strategy for Sustainable and Smart Mobility Before 2020–2021* (2020), https://www.europarl.europa.eu/legislative-train/theme-a-european-green-deal/file-sustainable-and-smart-mobility#:~:text=In%20the%20European%20Green%20Deal,to%20their%20current%20mobility%20habits. Accessed 20 June 2020

11. D. Tsokolis, A. Dimaratos, Z. Samaras, S. Tsiakmakis, G. Fontaras, B. Ciuffo, Quantification of the effect of WLTP introduction on passenger cars $CO_2$ emissions. J. Earth Sci. Geotech. Eng. **7**, 191–214 (2017)

12. J. Demuynck, D. Bosteels, F. Bunar et al., Diesel passenger car with ultra-low NOx emissions in real driving conditions. MTZ Worldwide **81**, 40–43 (2020)

13. Transport & Environment, *EU Must Withdraw Carmakers 'Licence to Pollute' as Data Shows New Cars Meet Limits* (2019), https://www.transportenvironment.org/newsroom/blog/eu-must-withdraw-carmakers%E2%80%99-%E2%80%98license-pollute%E2%80%99-data-shows-new-cars-meet-limits. Accessed 20 May 2020

14. J. Andersson, B. Giechaskiel, R. Muñoz-Bueno, E. Sanbach, P. Dilara, *Particle Measurement Programme (PMP) Light-Duty Inter-laboratory Correlation Exercise (ILCE_LD)*. Final report. Institute for Environment and Sustainability (2007), https://publications.jrc.ec.europa.eu/repository/bitstream/JRC37386/7386%20-%20PMP_LD_final.pdf. Accessed 17 May 2020

15. B. Giechaskiel, T. Lähde, Y. Drossinos, Regulating particle number measurements from the tailpipe of light-duty vehicles: the next step? Environ. Res. **172**, 1–9 (2019)

16. A. Krajinska, J. Müller, et al., New diesels, new problems. Report. *Transport & Environment* (2020), https://www.transportenvironment.org/sites/te/files/publications/2020_01_New_diesels_new_problems_full_report.pdf. Accessed 3 May 2020

17. B. Giechaskiel, Particle number emission of a diesel vehicle during and between regeneration events. Catalysts **10**, 587 (2020)

18. D. OudeNijeweme, P. Freeland, M. Behringer, P. Aleiferis, *Developing Low Gasoline Particulate Emission Engines Through Improved Fuel Delivery*. SAE Technical Paper (2014)

19. A. Suarez-Bertoa, V. Valverde, M. Clairotte, J. Pavlovic, B. Giechaskiel, V. Franco, Z. Krager, C. Astorga, On-road emissions of passenger cars beyond the boundary conditions of the real-driving emissions test. Environ. Res. **176** (2019)

20. CLOVE, *Study on post-Euro 6/VI Emission Standards in Europe, Progress in Task 2.2: Development of a New Array of Tests*. Presentation to the Advisory Group on Vehicle Emission Standards (AGVES) (2019), https://circabc.europa.eu/sd/a/a108e064-c487-4bf6-bb46-7faac76f8205/Post-EURO%206%20WT2.2_AGVES_2019_10_18%20V4.pdf. Accessed 14 June 2020

21. J. Andersson, C. Haisch, S. Hausberger, et al., *Measuring Automotive Exhaust Particles Down to 10 nm*. DownToTen (2020), https://cordis.europa.eu/project/id/724085/results. Accessed 10 May 2020

22. B. Giechaskiel, T. Lahde, R. Suarez-Bertoa, M. Clairotte, T. Grigoratos, A. Zardini, A. Perujo, G. Martini, Particle number measurement in the European Legislation and future JRC activities. Combust. Eng. **57**, 3–16 (2018)

23. https://wiki.unece.org/display/trans/PMP+Web+Conference+11+May+2020minutes

24. C.D. DiGiulio, J.A. Pihl, J.E. Parks II, M.D. Amiridis, T.J. Toops, Passive-ammonia selective catalytic reduction (SCR): understanding NH3 formation over close-coupled three-way catalysts (TWC). Catal. Today **231**, 33–45 (2014)

25. C. Livingston, P. Rieger, A. Winer, Ammonia emissions from a representative in-use fleet of light and medium-duty vehicles in the California South Coast Air Basin. Atmos. Environ. **43**, 3326–3333 (2009)
26. S.N. Behera, M. Sharma, Investigating the potential role of ammonia in ion chemistry of fine particulate matter formation for an urban environment. Sci. Total Environ. **408**, 3569–3575 (2010)
27. R. Suarez-Bertoa, M. Pechout, M. Vojtíšek, C. Astorga, Regulated and non-regulated emissions from Euro 6 diesel, gasoline and CNG vehicles under Real World driving conditions. Atmosphere **11**, 204 (2020)
28. T. Stocker, *The Physical Science Basis: Working Group I Contribution to the Fifth Assessment Report of the Intergovernmental Panel on Climate Change* (Cambridge University Press, New York, NY, USA, 2013). ISBN 978-1-107-05799-9
29. M.C. Sarofim, S.T. Waldhoff, S.C. Anenberg, Valuing the ozone-related health benefits of methane emission controls. Environ. Resour. Econ. **66**, 45–63 (2017)
30. US EPA, *Integrated Science Assessment for Ozone and Related Photochemical Oxidants.* Report (2020), https://cfpub.epa.gov/ncea/isa/recordisplay.cfm?deid=348522. Accessed 28 May 2020
31. A. Takigawa, A. Matsunami, N. Arai, Methane emission from automobile equipped with three-way catalytic converter while driving. Energy **30**(2–4), 461–473 (2005)
32. R. Saurez-Bertoa, A.A. Zardini, H. Keuken, C. Astorga, Impact of ethanol containing gasoline blend on emissions from a flex-fuel vehicle tested over the Worldwide Harmonized Light duty Test Cycle (WLTC). *Fuel* **143**, 173–182 (2015)
33. European Commission, *Establishing a Framework for Achieving Climate Neutrality and Amending Regulation (EU) 2018/1999* (European Climate Law). Proposal for a regulation of the European Parliament and of the Council, COM (2020) 80 final (2020). https://eur-lex.europa.eu/legal-content/EN/TXT/?qid=1588581905912&uri=CELEX:52020PC0080. Accessed 28 May 2020
34. World Health Organisation, *IARC Monographs on the Evaluation of Carcinogenic Risks to Humans*, vol. 88 (International Agency for Research on Cancer, 2006), https://monographs.iarc.fr/wp-content/uploads/2018/06/mono88.pdf. Accessed 1 June 2020
35. European Renewable Ethanol, *Overview of Biofuel Policies and Markets Across the EU-28.* Report (2018), https://www.epure.org/media/1738/epure-overview-of-biofuels-polices-and-markets-across-the-eu-28-2018-update.pdf. Accessed 01 June 2020
36. S. Han, H. Bian, Y. Feng, A. Liu, X. Li, F. Zeng, X. Zhang, Analysis of the relationship between $O_3$, NO and $NO_2$ in Tianjin, China. Aerosol Air Qual. Res. **11**, 128–139 (2011)
37. D.C. Carslaw, N.J. Farren, A.R. Vaughan, W.S. Drysdale, S. Young, J.D. Lee, The diminishing importance of nitrogen dioxide emissions from road vehicle exhaust. Atmos. Environ. X 1 (2019)
38. R. Suarez-Bertoa, M. Clairotte, A. Bertold, S. Nakatani, L. Hill, K. Winkler, K. Charlotte, K. Thorsten, Z. Rens, H. Boertien, C. Astorga, Intercomparison of ethanol, formaldehyde and acetaldehyde measurements from a flex-fuel vehicle exhaust during the WLTC. Fuel **203**, 330–340 (2017)
39. A. Ramos, J. Muñoz, F. Andrés, O. Armas, NOx emissions from diesel light duty vehicle tested under NEDC and real-word driving conditions. Transp. Res. Part D: Transp. Environ. **63**, 37–48 (2018)
40. D. Blanco-Rodriguez, G. Vagnoni, B. Holderbaum, EU6 C-Segment Diesel vehicles, a challenging segment to meet RDE and WLTP requirements. IFAC-*PapersOnLine* **49**, 649–656 (2016)
41. V. Valverde, B. Giechaskiel, M. Carriero, *Real Driving Emissions: 2019–2019 Assessment of Portable Emissions Measurement Systems (PEMS) Measurement Uncertainty.* JRC Technical Report (2020), https://publications.jrc.ec.europa.eu/repository/bitstream/JRC114416/jrc_pems_margin_review_nox_final_-_online_version.pdf. Accessed 25 June 2020
42. B. Giechaskiel, V. Valverde, Assessment of gaseous and particulate emissions of a Euro 6d-temp diesel vehicle driven >1300 km including six diesel particulate filter regenerations. Atmosphere **11**(6), 645 (2020)

43. M. Leblanc, L. Noël, B.R. Mili, B. D'Anna, S. Raux, Impact of engine warm-up and DPF active regeneration on regulated & unregulated emissions of a Euro 6 diesel SCR equipped vehicle. J. Earth Sci. Geotech. Eng. **4**, 29–50 (2016)

44. CLOVE, *Study on post-Euro 6/VI Emission Standards in Europe, Progress in Task 2.2: Development of a New Array of Tests.* Presentation to the Advisory Group on Vehicle Emission Standards (AGVES) (2019), https://circabc.europa.eu/sd/a/a108e064-c487-4bf6-bb46-7faac76f8 205/Post-EURO%206%20WT2.2_AGVES_2019_10_18%20V4.pdf. Accessed 14 May 2020

45. Transport & Environment, Groupe PSA Real world fuel economy measurements. https://www.transportenvironment.org/press/real-world-fuel-consumption-test-protocol-developed-groupe-psa-te-fne-and-bureau-veritas. Accessed 10th of April 2020 (2017).

46. B. Giechaskiel, V. Valverde, Assessment of gaseous and particulate emissions of a Euro 6d-temp diesel vehicle driven 1300 km including six diesel particulate filter regenerations. Atmosphere **11**, 645 (2020)

47. Z. Samaras, G. Mellios, Emission and fuel consumption tests. *Report for Transport & Environment* (2018), https://www.transportenvironment.org/sites/te/files/publications/Emisia% E2%80%99s%20testing%20report%20-%20Emissions%20and%20fuel%20consumption% 20tests.pdf. Accessed 01 June 2020

48. P. Van Mensch, R.F.A. Cuelenaere, N.E. Ligterink, *Assessment of Risk for Elevated NOx Emissions of Diesel Vehicles Outside the Boundaries of RDE: Identifying Relevant Driving and Vehicle Conditions and Possible Abatement Measures.* Report (TNO, 2017), https://repository. tudelft.nl/view/tno/uuid%3Ab0ff9bd6-41d0-4d88-89fe-012321e955be. Accessed 1 June 2020

49. F. Cuenot, *Real World Fuel Economy Measurements: Technical Insight from 400 Tests of Peugeot, Citroen and DS Cars.* Report, https://www.transportenvironment.org/sites/te/files/pub lications/Protocol_Technical_Insights%20%281%29.pdf. Accessed 1 June 2020

50. The European Automobile Manufacturer's Association, *Automobile Industry Pocket Guide 2019–2020* (2019), https://www.acea.be/publications/article/acea-pocket-guide. Accessed 5 June 2020

# Transportation Emissions on the Evolving European Agenda

Thomas Brewer

**Abstract** Transportation emissions account for one-fourth of Europe's greenhouse gas emissions, and they pose serious local public health problems in all European countries. Transportation's black carbon emissions—which are particulate matter—are also major contributors to climate change as well as health problems. Although motor vehicles are the primary source of emissions within the transportation sector, the emissions from maritime shipping, aviation and railroads also contribute to climate change and public health problems. EU elections in 2019 changed the policy agenda for the European Parliament and the Commission, as their new members in both were more supportive of action on climate change and other environmental issues. Action on transportation, furthermore, has been among the priorities on the lists of sectors needing action. However, there have been policy conflicts among member states, among party coalitions in the Parliament and among members of the Commission. This chapter focuses on specific policy issues concerning emissions in the motor vehicle, maritime shipping, aviation and railroad modes, and it analyzes them in the context of pandemic-induced economic recovery programs.

## 1 Introduction

In early 2020, as the newly installed EU Commission and Parliament were focusing increasing attention on their sustainable development agendas, including transportation emissions, in particular, the coronavirus pandemic suddenly became the dominant issue in Brussels. Yet, by mid 2020, as this chapter was being completed, transportation emissions and other sustainability issues had re-emerged on the active EU agenda.

In Europe, transportation emissions contribute one-fourth of the total greenhouse *gas* emissions that cause climate change [1]. In addition, black carbon *particulates* are a significant cause of both climate change and public health problems. Motor

T. Brewer (✉)
Georgetown University, Washington, DC, USA
e-mail: brewert@georgetown.edu

vehicles are the primary modal source of the emissions within the transportation sector; they contribute more than 70% of the transportation sector's greenhouse gas emissions [1]. However, emissions from maritime shipping, aviation and railroads are also significant contributors to climate change and public health problems, especially in large urban areas.

The relatively high rates of emissions from motor vehicles and aviation are evident in the data of Table 1. In that table, the comparative emissions data for the amount of $CO_2$ per passenger, per kilometer reveal clear patterns in the relative ratios across modes, with rail being the common base for all the comparisons. Maritime shipping is not included in the table because their freight emission data are not comparable with the passenger data in the table.

Aviation emissions in the EU and UK are much greater than rail emissions per passenger, per kilometer—approximately 20 times greater. Automobiles with one passenger or one-and-a-half on average are also relatively high—indeed nearly 30 times higher than rail in the UK study and 7–11 times higher in the EU study. Automobiles with four passengers inevitably have lower emissions per passenger, but still much higher per passenger than rail.

There are also of course differences within each of the four modes. The differences among types of train service are especially pertinent because they indicate how much emissions can still be reduced within that mode. The 14 g per passenger per kilometer in the EU study is a useful anchoring point for two comparisons from the UK. Domestic trains in the UK, on average, are about three times greater than the EU average. The high-speed Eurostar trains running between the UK and continental

**Table 1** European transportation emissions: comparisons among modes [2, 3]

| Mode | $CO_2$ Emissions—grams/per passenger/per kilometer (ratio = highest/lowest emission level in the column) | |
|------|------|------|
| | EU | UK |
| Aviation (88 passengers) | 285 (20.4) | |
| Aviation—Domestic (UK) | | 133 (22.1) |
| Aviation—Long (UK Intl.) | | 102 (17.0) |
| Auto—SUV, 1.5 passengers | 158 (11.3) | |
| Auto—1.5 passenger | 104 (7.4) | |
| Auto—1 passenger | | 171 [diesel] (28.5) |
| Auto—SUV 4 passengers | 55 (3.9) | |
| Auto—4 passengers | 42 (3.0) | 43 [diesel] (7.2) |
| Bus—12.7 passengers | 68 (4.9) | |
| Bus | | 104 (17.3) |
| Rail (156 passengers) | **14** | |
| Rail—Domestic (UK) | | 41 (6.8) |
| Rail—Eurostar (UK-Fr) | | **6** |

Europe, on the other hand, are less than half the EU average. Thus, modern high-speed rail service among major population centers offers many opportunities for reducing emissions—an opportunity that is analyzed and illustrated in the section below on modal policy issues.

How Europe responds to the transportation emission issues addressed here and in the other chapters will be decided in a complex array of policymaking processes at regional, national and local governance levels. An already complex array of multi-level policymaking processes was compounded by the coronavirus pandemic and ensuing economic recession. The remainder of this chapter identifies key issues, proposals and decisions that were 'in play' in mid 2020.

## 2   Economic Recovery Policies

There are significant pandemic economic recovery policies at both the EU level and the national government level. There have been three policymaking tracks where there have been large-scale initiatives: the European Central Bank (ECB), the EU Commission and national governments. In early May 2020, the ECB announced a bond-buying program of 750 billion euros named the Pandemic Emergency Purchase Program (PEPP), and it then added another 600 billion in early June for a total of 1.35 trillion euros in order stimulate the EU economy [4]. In early June, the ECB was forecasting an 8.7% decline in 2020 GDP and a price increase of only 0.3%, and it was the prospect of a decline into a deflationary period that prompted it to act. Although its program had no direct relevance to funding transportation projects or regulating transportation emissions, it was a strong action that encouraged more optimism about the state of the EU economy over the next few years.

In mid-May, Chancellor Merkel and President Macron made a joint proposal for a 500 billion euro fund that would make grants to member governments with funds from EU bonds sold in financial markets [5]. Four EU members—Austria, Denmark, the Netherlands and Sweden, who came to be known as the 'frugal four'—proposed an alternative: loans instead of grants, funding through savings in other programs instead of EU bonds and restrictions on the timing and conditionality of the loans [6].

Then on 23 May, the Commission proposed a 750 billion euro plan with 500 billion in grants and 250 billion in loans [7]. The Commission further proposed that the programs be financed by a combination of new taxes, reforms of the Emissions Trading System (ETS), border adjustment measures (BAMs) on carbon-intensive imports and a tax on large information technology firms. The Commission programs would need to be incorporated in the next seven-year budget beginning in 2021. There were many budget 'details' to be negotiated, including how the funds would be distributed and how they would relate to programs in the regular 2021–28 budgets. Early estimates of the separate French and German national government programs were that the total French economic support would be about 88 billion euros and Germany's about 457 billion euros.

In the context of this book's focus on transportation emissions, of course key questions were: How much of the money would be for reducing those emissions? How much of it would lead to increased air-polluting emissions? Answers to these questions would depend on the fate of the European Green Deal [8], which the newly elected Commission proposed in December 2019, and its Transportation Strategy [9].

## 3   The Transportation Strategy and the European Green Deal

The Transportation Strategy adopted by the Commission and agreed by the Parliament and the Council provides a basis for anticipating the outlines of future transportation policies. A concise summary of the strategy is that it falls within the 'sustainable mobility' focus of EU long-term planning, and it 'aims at promoting and rolling out more sustainable – cleaner, cheaper and healthier – forms of private and public transport' [10]. More specifically, it provides as indicated in the box below:

**Box: Transportation Sector in the EU Strategic Plan [7]**

The 'main elements' of the strategy for 'low-emission mobility,' as announced in 2016, are:

- Increasing the efficiency of the transport system by making the most of digital technologies, smart pricing and further encouraging the shift to lower emission transport modes.
- Speeding up the deployment of low-emission alternative energy for transport, such as advanced biofuels, electricity, hydrogen and renewable synthetic fuels and removing obstacles to the electrification of transport.
- Moving toward zero-emission vehicles. While further improvements to the internal combustion [engine] will be needed, Europe needs to accelerate the transition towards low- and zero-emission vehicles.

The strategy notes that 'cities and local authorities are crucial for the delivery of this strategy.' It also reiterates a commitment to action at the global level to reduce emissions in international aviation and international shipping. (In fact, both the International Civil Aviation Organization (ICAO) and the International Maritime Organization (IMO) have been put on notice that if there is not sufficient progress toward international emission trading systems in those industries, the EU will take regulatory action on its own.)

The strategy also notes that it 'draws upon existing mechanisms and funds'—including 39 billion euros available at that time in funds 'for supporting the move toward low-emission mobility' in the structural investment fund and 6.4 billion euros for low-carbon mobility projects in the Horizon 2020 program.

These statements of emphasis and intent take on greater significance as the EU undertakes policy initiatives by the new Commission and in light of the European Green Deal [11]. In a 2020 speech to the Parliament, Commissioner for Transport Valean said the 'Strategy on Sustainable and Smart Mobility' will have 'these two objectives at its very heart' (where 'smart' implies 'digital'). She further enunciated 'four areas of action' where transportation could make contributions to the European Green Deal, as noted in the box below.

**Box: EU Transportation Preliminary Proposals for Action [12]**
- Boost[ing] the uptake of clean vehicles and alternative fuels for road, maritime and aviation.
- Increasing the share of more sustainable transport modes such as rail and inland waterways and improving efficiency across the whole transport system.
- Incentivizing the right consumer choices and low-emission practices.
- Investing in low- and zero-emissions solutions, including infrastructure.

The European Green Deal, of course, includes many economic sectors and industries other than transportation, and the emissions, technologies and policies interact with those in the transportation sector. An obvious example is the importance of wind, solar and other sustainable sources of electricity to all modes of transportation—with the deployment of electric recharging stations for motor vehicles and ships being two timely illustrations.

At the same time, each mode poses its own distinctive challenges, as indicated in the following sections.

## 3.1 Railroads

The only mode of public transportation in Europe that has increased its business while reducing its emissions is railroads. It has accomplished this by increasing its dependence on electricity and at the same time reducing its dependence on coal-fed sources of electricity. In the Netherlands and Denmark, trains predominantly use wind energy [13].

Recent research about consumers and the plans of governments and railroads indicated that railroad traffic should increase in coming years [13]. Passengers are increasingly likely to take a train instead of a plane for trips of 4 h or less. New high-speed trains are being planned, for example, between Dresden and Prague, thereby reducing the travel time from 4 1/2 to 2 1/2 h. A tangible example of national government policy incentivizing a shift to rail from aviation is the French government plan announced by President Macron [14]. The plan includes an economic recovery package of 7 billion euros for Air France-KLM. Among the conditions is that the airline must terminate relatively short flights between cities served by high-speed trains—presumably including Paris–Brussels and Paris–London.

On the political marketing front, the EC has indicated a desire to proclaim 2021 the 'year of rail' [15]. The Green Deal includes a proposed goal of reducing the 75% of freight that is currently transported by trucks to both rail and marine shipping within the EU [8]. A group of 23 EU members plus non-members Norway and Switzerland agreed to a declaration in June 2020 that the Green Deal should include an effort to improve the international passenger rail services within Europe [16].

It should be noted, though, that of course trains are not entirely emission-free. There are significant differences between diesel and electric locomotives, with the former emitting as much as twice the latter, depending on the source of the electricity. Paddington passenger train station in London—where neither indoor nor outdoor air quality regulations apply in the covered platform areas—has had concentration levels of PM, $NO_2$ and $SO_2$ in the station greater than those in the streets outside the station and above legal limits for outdoor locations [17]. However, there is a technological solution, namely diesel particulate filters (DPFs) [18].

In sum, railroads in Europe are likely to get more attention, more funding and more regulations, and they will grow over the next several years [13]. While rail is in ascendance in both industry expectations and governmental policies, aviation is expected to be in relative descent, compared with its situation before the pandemic [19].

## 3.2 Aviation

In some respects, aviation poses especially problematic challenges to governments. Airlines are high-profile firms in their home countries, and in some countries, airplane manufacturing is also high profile (e.g., Airbus in France, Germany and Spain). There are therefore symbolic political as well as real economic benefits to incentivize governments to come to the corporations' aid. Yet, the industry's losses have been so large during the pandemic and recession that bailouts of the airlines and manufacturers are enormously expensive.

Some of the bailout subsidies, however, have had conditions attached to them. Highlights of the French government's package are in the following box:

---

**Box: French Plan to Aid Its Aviation Industry [14]**
- The plan includes 15 billion euros total for both the airline Air France-KLM and the airplane manufacturer Airbus and other firms in the manufacturing supply chain.
- There is a combination of 3 billion in loans from the government and 4 billion in loan guarantees for a sub-total of 7 billion euros for Air France-KLM.

- There is a condition that the airline end short-distance flights between cities that are served by high-speed rail. Details of the routes and schedules were not yet public when the plan was announced.
- The plan for the manufacturer Airbus and other parts of the supply chain are more complex—with a sub-total of 8 billion.
- Most of the 8 billion will go to Airbus, but with other recipients as well.
- There is a special purpose carbon reduction program that amounts to a form of conditionality—namely 1.5 billion euros is being offered over three years to improve the sustainability of airplanes—with a goal of a carbon-neutral plane by 2035.
- An investment fund is also being created to help small- and medium-sized firms in the supply chain. It would begin with 500 million euros and increase to twice that amount. Of the 500 million initial funding, 200 is from the government, 200 from industry and 100 from the fund manager.

The highlights of the German package are below. An important condition imposed by the EU Competition Policy Directorate is that Lufthansa has to give up some of its slots at the Frankfurt and Munich airports, where it had about two-thirds of the slots in each one. Although Lufthansa and the German government initially objected to this Commission condition, they eventually accepted it.

**Box: German Plan to Aid Its Aviation Industry [20]**
Germany's stimulus plan for its aviation industry offered:
- A total financial plan of 9 billion for Lufthansa.
- The government will buy a 20% share in Lufthansa for 300 million euros.
- The government development bank will lend Lufthansa 3 billion euros.
- The government will have two seats on Lufthansa's board.
- Lufthansa will not pay dividends and it will limit executives' pay.

Yet, many governments have agreed to major bailouts without any conditions. According to one count, there were more than 20 billion euros already agreed with no conditions and another 8 billion still being discussed [21]. One policy change that could come under much closer scrutiny is to eliminate the tax free status of airline jet fuel [22].

European policies at the International Civil Aviation Organization (ICAO) were also in doubt as the individual European countries, including the 27 in the EU, reviewed their policies in light of the pandemic, the recession and evolving EU-level policies [23]. The EU is not a formal member among the 193 national governments that are, but the EU is one of the more than 70 'invited organizations' at the ICAO. All 27 of the EU members are ICAO members, as are Iceland, Monaco, Norway, Switzerland and the UK [24].

## 3.3 Maritime Shipping

International shipping is often divided into many 'classes' such as cruise ships and many kinds of freight classes: container, tankers, dry bulk and others. The various classes exhibit a combination of their common and distinctive emissions issues.

The cruise segment was already being scrutinised more thoroughly, and in some countries, it was being more directly regulated before the pandemic. These tendencies were evident in both northern and southern Europe. In the north, there was increasing interest among Baltic and North Sea countries in coordinating seaport policies in order to avoid cruise ships taking advantage of differences in regulations. At the same time, there was increasing interest in developing ports in order to be more accommodating to cruise ships, thus attracting more business on shore. The pandemic health crisis and the ensuring economic recession, however, put such plans on hold. For instance, the Copenhagen Malmo Port (CMP), which planned a major expansion to be completed in 2022, decided in the spring of 2020 to wait for a clearer picture of the future of the industry [25].

In the south, there was increasing interest in the formation of a Mediterranean Emission Control Area [26]. There is a strong pattern of north–south routes, where cruise ships carry European tourists to North African countries (with some intra-European east–west traffic as well), and east–west routes, where goods from China are passing into the Mediterranean to ports in France, Greece, Italy, Spain and also beyond to the UK, the Netherlands, Germany and the Nordic countries. Thus, there is a mix of economic and political issues, with patterns along geographic lines. The north–south cruise routes are significant economic issues for the North African countries with tourist centers along the south Mediterranean coast. For their European home ports, the cruise ships' air pollution emissions are a sensitive issue, in Barcelona, for instance, where the emissions are especially dense. The west–east routes for container ships and other types of goods-carrying ships, include only a few Mediterranean ports, for instance, in Greece and Italy, where there are major economic issues; otherwise, it is ports beyond the Mediterranean where the goods are headed where there are economic issues. For this trade, the Mediterranean is mostly a convenient route, where ships emit pollutants that end up in both European and North African coastal areas, with coastal populations totaling in the hundreds of millions.

Globally, for the container segment of international maritime shipping, a key issue for the post-pandemic era was whether many firms would change their international production and supply chain strategies. A combination of more geographically diversified manufacturing and supply sources, together with regionalization, would provide greater security against future pandemics [27].

Meanwhile, intra-coastal maritime shipping within Europe was already getting increased attention in 2020-for instance, a consortium named Zero-Emission Services included shipbuilder Wartsila, ING bank, technical services firm Engie and the Port of Rotterdam, with the support of the Dutch Ministry of Infrastructure and Water Management. Their plan is to replace diesel-powered barges and other ships with electricity-powered vessels. Such a change of technologies could significantly reduce the Netherlands' $CO_2$ emissions [28].

These contemporary issues about the future of maritime shipping emerge in the context of well established but still evolving EU policies dating back to 2008, when the Marine Strategy Framework Directive was adopted by the Commission [29]. The directive listed four regions: Baltic, North-west Atlantic, Mediterranean, and Black. It included targets for 2020 and measures to achieve the targets, and monitoring processes to gauge progress. An ambitious objective was to have a 'zero-emission' maritime shipping industry. A review in 2016 found that there had been progress in several respects [30]. Among the highlighted developments were the IMO adoption of a new Energy Efficiency Design Index and an Energy Efficiency Management Plan, the development of an EU $CO_2$ monitoring, reporting and verification (MRV) regulation applying to all ships entering EU ports beginning January 2018, reduction of the sulfur content of ships' fuel in European Sulpher Emission Control Areas (SECAs) effective January 2015 and the IMO adoption of new worldwide fuel sulfur content limits effective in January 2020. The central issue in 2020, therefore, was whether this decade-long momentum would be extended into the 2020s.

Of course, the future of emission regulations depends not only on EU institutional policymaking but also local port authorities and the International Maritime Organization (IMO) at the global level. At the IMO, the European Commission has a Cooperation Agreement as an International Organization, but active de facto representation of European interests in the IMO decision-making process is by the 27 EU member states, plus Iceland, Monaco, Norway, Switzerland and the UK [31]. There is a degree of cooperation among the EU member states, with Nordic governments often in the lead.

At the local level, a report by the International Transport Forum (ITF) at the OECD projected worldwide port increases of 100–400% by 2050 in six types of emissions [32]. However, European ports were expected to produce only 5% of world SOx emissions, for example, though European ports were also expected to process about one-fourth of the world total shipping traffic. Some European ports have already been actively involved in developing emission control measures. Several ports have installed shore-to-ship electric power systems so that ships do not need to run their diesel-powered auxiliary engines while in port.

## 3.4 Motor Vehicles

There is a strong and expanding movement in Europe to phase-out combustion engine automobiles—both diesel and gasoline. At least nine national governments, two regional governments and eight city governments had announced specific plans by April 2020, with phase-out years varying from 2024 to 2040. Table 2 lists the countries and regional or city governments within the countries that have their own plans. (Milan and Rome have city plans, but there was no national plan for Italy as of April 2020. In Belgium, there was no national plan, but there is a Brussels region plan).

In late 2019, Denmark and other EU member states put the issue of an EU-wide phase-out of diesel and gasoline cars on the agenda of the Commission and the Parliament [34], and there is likely to be increasing interest in the issue as more national governments and cities take action on air pollution in the aftermath of the coronavirus.

At the national level, France's announcement in May 2020 and Germany's announcement in June of their financial assistance programs for their automotive industries are described in the following boxes.

**Table 2** European countries with plans to phase-out combustion engine cars[a] [33]

| Country<br>Target year | Key provision of plan |
|---|---|
| Denmark<br>2030 | No new diesel or gasoline cars will be sold—and no PHEVs[b] (2035) |
| France<br>2040 | No new fossil fuel passenger cars or light commercial vehicles will be sold<br>Paris and Strasbourg: additional restrictions |
| Iceland[e]<br>2030 | No new registrations of diesel or gasoline cars |
| Ireland<br>2030 | No new sales of fossil fuel cars |
| Netherlands<br>2030 | New passenger cars will be emission-free |
| Norway<br>2025 | All new passenger cars and light vans will be zero-emission<br>Bergen and Oslo: additional restrictions |
| Slovenia<br>2030 | New passenger cars will be emission-free |
| Spain<br>2040 | 100% of passenger cars sold will be electric |
| Sweden<br>2030 | No new diesel or gasoline cars will be sold |
| United Kingdom[e]<br>2035 | No new diesel or gasoline cars or vans, or PHEVs[b] or HEVs[c] will be sold<br>Scotland: phase-out diesel and gasoline cars and vans (2032)<br>London: phase-out fossil fuel vehicles, establish central ZEZ[d] (2025) |

[a] As of April 2020
[b] PHEV—Plug-in Hybrid Electric Vehicle
[c] HEV: Hybrid Electric Vehicle
[d] ZEZ: Zero-Emission Zone
[e] Not in EU

**Box: French Plan to Aid Its Auto Industry [14]**

France's auto plan includes:

- 5 billion euros in assistance for Renault and 3 billion for PSA, which makes Peugeot and Citroen cars.
- The national production target is 1 million 'clean' cars per year by 2025; 'clean' cars are electric and hybrid and some already-produced gasoline and diesel cars. PSA, which produced no 'clean' cars in 2019, is expected to produce 450,000 a year by 2025; Renault, which is 15% owned by the government, is expected to triple its current output of 'clean' cars by 2022.
- There will be subsidies of as much as 12,000 euros for purchasers of electric cars.
- Another target is to have 100,000 charging stations by the end of 2021.
- In addition, a long-term strategic objective is to 'relocalize' production in France. Toward this end, a new project is to produce batteries for electric vehicles. There will be three partners in the project: the government, PSA and Saft, which is a subsidiary of the French oil firm Total.

**Box: German Plan to Aid Its Auto Industry [36, 37]**

Germany's stimulus plan for its auto industry offered:

- Subsidies of 6000 euros to buyers of new electric vehicles.
- A temporary 3% reduction of the Value Added Tax (VAT) on purchases of some autos.
- An 'innovation premium' for auto manufacturers, with 2 billion euros for hybrid autos.
- It also included 2.5 billion euros for electric vehicle recharging stations and 1.2 billion euros for less polluting busses and trucks.
- However, it did not offer a large-scale scrappage scheme, as it had to the extent of 5 billion euros during the financial crisis a decade previously.

## 4  Conclusion: Paths into the Future

In mid 2020, as many proposals for Europe's economic recovery policies and projects were being made public, there was a proposal for 'breaking away from the path dependency' that could retard technological and policy progress [38]. There were three transportation-related proposals:

- The airline industry should 'reinvent itself and become a travel industry, not just an airline industry. Short-haul flights could be replaced with high-speed trains, which could also bring passengers to major airports.'
- 'Intra-European rail transport must be considered as a matter of public service—tariffs need to be regulated, and one should be able to buy a single ticket to cover the whole journey.'
- 'While intra-urban commuters are likely to bike more, suburban commuters are likely to drive more, with devastating consequences on GHG emissions. Bike lanes must therefore be built for longer commutes from the suburbs to the inner cities.'

Such proposals may be well received in Brussels and elsewhere in Europe. For there is a widely shared aspiration in Europe to continue to be world leaders in policies and technologies to reduce emissions that contribute to climate change and other problems [39].

In fact, there is also a documented record of accomplishments to date in that regard. For instance, results of the 2020 Environmental Performance Index (EPI) provide ample evidence of European accomplishments.

Wendling et al. [40] provide extensive empirical support for this record. Based on rankings of 180 countries, the top ten places were all awarded to European countries—with Denmark first and their Nordic neighbors Finland, Sweden and Norway all in the top ten. Germany, France and the UK were also in the top ten, along with Austria, Luxembourg and Switzerland. One of its many specific indicators is PM2.5 exposure, which is based on the number of 'standardised disability-adjusted life-years lost per 100,000 persons … due to exposure to fine air particulate matter smaller than 2.5 $\mu$m.' Although not specifically about any particular transportation sources, it is a useful measure that includes black carbon ($PM_1$) and other pollutants from transportation. Of the top ten (lowest emissions) in the world, seven were European countries, and of the top 20, 15 were European countries.

Of course, such summary rankings based on the past are no guarantee of future progress on specific transportation emission issues, but they do implicitly suggest an interest in addressing transportation emission issues in the context of a broader commitment to action on air pollution issues. Moreover, the data and detailed examples of recent actions and announced plans for the near future strongly suggest that many individual national governments, as well as local governments and EU institutions, will be world leaders in limiting transportation emissions.

# Appendix

## List of 32 'European' Countries and their Institutional Affiliations

| Europe 32 | EU 27 | Euro 19 | EFTA 4 | UK |
|---|---|---|---|---|
| Austria | y | y | | |
| Belgium | y | y | | |
| Bulgaria | y | | | |
| Croatia | y | | | |
| Cyprus | y | y | | |
| Czechia | y | | | |
| Denmark | y | | | |
| Estonia | y | y | | |
| Finland | y | y | | |
| France | y | y | | |
| Germany | y | y | | |
| Greece | y | y | | |
| Hungary | y | | | |
| Iceland | | | y | |
| Ireland | y | y | | |
| Italy | y | y | | |
| Latvia | y | y | | |
| Liechtenstein | | | y | |
| Lithuania | y | y | | |
| Luxembourg | y | y | | |
| Malta | y | y | | |
| Netherlands | y | y | | |
| Norway | | | y | |
| Poland | y | | | |
| Portugal | y | y | | |
| Romania | y | | | |
| Slovakia | y | y | | |
| Slovenia | y | y | | |
| Spain | y | y | | |
| Sweden | y | | | |
| Switzerland | | | y | |
| UK | | | | y |

# References

1. European Environment Agency (EEA), *The European Environment—State and Outlook 2020* (2019), https://www.eea.europa.eu/. Accessed 27 June 2020
2. European Environment Agency (EEA), *Transport and Environment Report 2019: The First and Last Mile—The Key to Sustainable Urban Transport* (2020), https://www.eea.europa.eu/publications/the-first-and-last-mile. Accessed 3 June 2020
3. R. Blunt, How to travel by train—and ditch the plane. *British Broadcasting Corporation (BBC)* (2020, January 7), https://www.bbc.com/news/world-europe-51007504. Accessed 25 June 2020
4. M. Arnold, ECB boosts bond-buying stimulus package by [600 billion euros]. *Financial Times* (2020, June 4), https://www.ft.com. Accessed 4 June 2020
5. F. Giugliano, Merkel and Macron make a stunning coronavirus proposal. *Bloomberg* (2020, May 18), https://www.bloomberg.com. Accessed on 20 May 2020
6. EURACTIV, *'Frugal Four' Present Counter-Plan to Macron-Merkel EU Recovery Scheme* (2020), www.euractiv.com. Accessed 2 June 2020
7. European Commission, *Commission Publishes Strategy for Low-Emission Mobility* (2016), https://ec.europa.eu. Accessed 28 May 2020
8. European Commission, *The European Green Deal* (Brussels, 2019). Accessed 28 May 2020
9. European Commission, *Climate Action—Transport Emissions: A European Strategy for Low-Emission Mobility* (2020), https://ec.europa.eu/clima/policies/transport_en. Accessed 25 Jan 2020
10. Florence School of Regulation, *The European Green Deal* (2020), https://fsr.eui.eu/the-european-green-deal/. Accessed 21 May 2020
11. European Commission, *The European Green Deal* (Brussels, 2019)
12. European Commission, *Commissioner Valean's Speech: EU Strategy for Mobility and Transport: Measures Needed by 2030 and Beyond* (2020), https://ec.europa.eu. Accessed 28 May 2020
13. International Energy Agency (IEA), *The Future of Rail* (2019), https://www.iea.org/. Accessed 3 June 2020
14. Financial Times, *France Unveils [15 billion euro] Aid Package to 'Save' Its Aerospace Industry* (2020), https://www.ft.com. Accessed 9 June 2020
15. European Commission, *Promoting Sustainable Mobility: Commission Proposes 2021 to be the European Year of Rail* (2020), https://ec.europa.eu/commission. Accessed 10 Mar 2020
16. EURACTIV, *24 Countries Sign Pledge to Boost International Rail Routes* (2020), https://euractiv.com. Accessed 6 June 2020
17. U. Chong et al., Air quality evaluation of London Paddington train station. Environ. Res. Lett. **10**(9), 1–12 (2015)
18. Manufacturers of Emission Controls Association (MECA), *Case Studies of the Use of Exhaust Emission Controls on Locomotives and Large Marine Diesel Engines* (2014), www.meca.org. Accessed on 3 June 2020
19. EURACTIV, *Planes vs. Trains: High-Speed Rail Set for Coronavirus Dividend* (2020), www.euractiv.com. Accessed 9 May 2020
20. Financial Times, *German Government Agrees [9 billion euro] Bailout for Lufthansa* (2020), https://www.ft.com. Accessed 25 May 2020
21. Transport & Environment, *Bailout Tracker* (2020), https://www.transportenvironment.org. Accessed 6 May 2020
22. J. Faber, A. O'Leary, *Taxing Aviation Fuels in the EU* (2018), www.cedelft.eu. Accessed 1 June 2020
23. International Civil Aviation Organization (ICAO), *Member States and Invited Organizations* (2019), https://www.icao.int/about-icao/.org Accessed 24 June 2020
24. Climate Change News, *Final Blow to Aviation Climate Plan as EU Agrees to Weaken Rules* (2020), https://www.climaechangenews.com. Accessed 12 June 2020

25. M. Latarche, Copenhagen's new cruise terminal postponed. *ShipInsight* (2020, June 1), https://shipinsight.com. Accessed 1 June 2020
26. T. Brewer, A maritime emission control area for the Mediterranean Sea? Technological solutions and policy options for a 'Med ECA.' Euro-Mediterr. J. Environ. Integr. **5**(15), 35–42 (2020)
27. Financial Times, *Pandemic Strains Shipping, Air and Rail Frieght Operators* (2020). https://www.ft.com. Accessed 6 October 2020
28. M. Latarche, *Wartsila and Partners Develop Emissions-Free Barge Concept*. ShipInsight (2020, June 3), https://shipinsight.com. Accessed 3 June 2020
29. European Commission, *Directive 2008/56/EC, Marine Strategy Framework Directive* (2008, June 17). https://ec.europa.eu. Accessed 7 June 2020
30. European Commission, *Reducing emissions from the shipping sector* (2016). https://ec.europa.eu. Accessed 6 October 2020
31. International Maritime Organization (IMO), *Member States* (2020), https://www.imo.org. Accessed 14 June 2020
32. International Transport Forum (ITF)/OECD, *Shipping Emissions on Ports*. Discussion paper 2014, 20, by Olaf Merk (Paris, 2014)
33. S. Wappelhorst, *The End of the Road? An Overview of Combustion-Engine Car Phase-Out Announcements Across Europe*. International Council on Clean Transportation (ICCT) (2020), https://theicct.org/publications/combustion-engine-car-phase-out-EU. Accessed 3 June 2020
34. E&E News, *Europe: Common Criteria Sought for Green State Aid* (2020), https://www.eenews.net/climaewire. Accessed 23 Apr 2020
35. V. Mallet, *Emmanuel Macron Injects [8 billion euros] to Fuel French Car Industry Revival* (2020), https://www.ft.com. Accessed 26 May 2020
36. EURACTIV, *Germany's [130 billion euro] Stimulus Packaged Praised by All Sides* (2020), https://eurative.com. Accessed 6 June 2020
37. E. Bannon, *Germany's Green Rescue Package Points the Way for Europe* (2020), https://tranportenvironment.org. Accessed 5 June 2020
38. B. Picard, *Breaking Away From the Path Dependency. EURACTIV* (2020, June 8), https://www.euractive.com. Accessed 8 June 2020
39. Finland, *The EU as a Global Leader in Climate Action* (2019), https://eu2019.fi. Accessed 22 July 2019
40. Z.A. Wendling, J.W. Emerson, A. de Sherbinin, D.C. Esty, et al., *Environmental Performance Index* (Yale Center for Environmental Law & Policy, New Haven, CT, 2020). https://epi.yale.edu/epi-results/2020/component/epi. Accessed 6 June 2020

# Emission Trading Systems in Transportation

**Elizabeth Zelljadt and Michael Mehling**

**Abstract** Emissions trading systems (ETSs) apply a market-oriented approach to the control of pollutant emissions, affording flexibility to emitters to decide when and where emissions will be abated. Most ETSs to date have applied to a limited number of stationary sources in industry and the power sector, where emissions can be easily monitored and the ETS itself more easily administered. Still, the appeal of emissions trading as a market-based policy instrument has also prompted their deployment to reduce emissions from the transportation sector, usually by including fuels upstream as these enter into the market. Following a short introduction to the concept of emissions trading, this chapter provides an overview of four case studies where emissions trading has been applied to transportation: the New Zealand ETS; the Western Climate Initiative; the Transport and Climate Initiative; and the Carbon Offsetting and Reduction Scheme for International Aviation. It concludes with a brief analysis of lessons learned and prospects for expanded use of emissions trading to manage emissions from transportation.

## 1 Introduction

Emissions trading systems (ETSs) are policy instruments that involve pollution quantity control coupled with a defined, tradable unit—such as a permit to emit a specified amount of a pollutant for a specified duration of time. They are often referred to as market mechanisms because they give rise to dynamics similar to those found in conventional markets for goods and services. They have been used to reduce emissions of sulphur dioxide and nitrogen oxides from power plants [1] and over the past two decades have become an important instrument for governments at all levels

E. Zelljadt
Carbon Team, Energy Commodities Content and Research, Refinitiv, Oslo, Norway
e-mail: ezelljadt@gmail.com

M. Mehling (✉)
MIT Center for Energy and Environmental Policy Research, Cambridge, MA, USA
e-mail: mmehling@mit.edu

© The Author(s), under exclusive license to Springer Nature Switzerland AG 2021
T. Brewer (ed.), *Transportation Air Pollutants*,
SpringerBriefs in Applied Sciences and Technology,
https://doi.org/10.1007/978-3-030-59691-0_7

(from regional to international) to achieve their greenhouse gas emission reduction targets.

Typically, regulators set a ceiling, or 'cap', on the amount of pollutants allowed to be emitted by a collection of covered entities and issue a correspondingly limited amount of permits to emit (usually called allowances) that can be traded—hence the name 'cap-and-trade' for this type of programme. As an alternative to 'command and control' style methods of addressing pollution, in which regulators set fixed quotas or limits on individual entities' pollution output and fine or punish them if those limits are exceeded, the market-based approach has found favour with regulators and those regulated for its flexibility: though the cap on emissions is fixed, exactly how the necessary reductions are achieved is determined by the market. Entities that are able to reduce emissions at low cost may do so beyond the extent required, selling their surplus allowances to entities for whom reduction is more expensive. Thus, the same net amount of reductions is achieved, but at the collective least cost.

More specifically, regulators avoid taxing pollution because they do not set a price: by incorporating tradable units, market mechanisms *reveal* an explicit price for environmental harm at the intersection of demand and supply, with the latter determined by the regulator. That price, in turn, internalises some or all of the social cost of environmental harm in the private cost of underlying behaviour, thereby helping to operationalise the polluter pays principle [2].

From an environmental perspective, the approach works best for pollutants like greenhouse gases that are harmful only at the macro (global) level—it does not work as well for pollutants that are toxic or otherwise harmful directly at the point of emission. Indeed, the case studies detailed here, in which transport emissions are part of an ETS, pertain to greenhouse gases rather than black carbon or criteria pollutants. From a regulatory perspective, the cap-and-trade approach typically works best when applied to a finite number of large stationary sources (such as factories and power plants) as their emissions can be more easily monitored at the source and reported for purposes of issuing allowances and tracking compliance.

The latter challenge constitutes one of the main reasons the transport sector, with its millions of vehicles from which emissions would have to be individually measured, has *not* been one to which cap-and-trade programmes have explicitly applied. However, some ETS have decided to incorporate emissions from vehicles by covering the fuels they combust 'upstream', i.e. before they are combusted in motors and emitted into the atmosphere as greenhouse gas. This chapter looks at these cases (New Zealand's national ETS and North America's Western Climate Initiative), as well as a planned ETS specifically targeting the motor vehicle transport sector (the Transport and Climate Initiative among north-eastern US states) and greenhouse gas emissions from international aviation (the International Civil Aviation Organisation's offsetting programme known by its acronym CORSIA).

## 2 New Zealand's ETS

The New Zealand Emissions Trading System (NZ ETS), in operation since 2008, was originally intended to cover all sectors of the island nation's economy, including not only transportation, but also agriculture and forestry as sources and sinks. Though agriculture is still not covered, the NZ ETS has the broadest sectoral coverage of any ETS in the world [3] due to its upstream point of regulation including transport fuel. In 2017, the transport sector accounted for one-fifth of New Zealand's gross emissions, over 90% of which come from road transport [4, 5].

Regulators applied the so-called point of obligation—the entity at which a compliance obligation exists in the fossil fuel use chain, from extraction of the fuel to combustion in a motor vehicle—very far *upstream*, meaning closer to the source of the fuel rather than to the point of emissions. The obligation is at the point where a liquid fossil fuel supplier imports fuel or takes fuel from a refinery. It was set there to minimise costs to ETS participants, lower the government's administration costs, and to enable effective monitoring and verification while still capturing as many emissions as practically possible and passing on effective economic incentives [6]. It made for 275 entities in 2018 [7] having a compliance obligation under the ETS, i.e. having to surrender allowances for the emissions associated with the fuel they sold. It applies to suppliers of all major liquid fuels used domestically: petrol, diesel, aviation gasoline, jet kerosene, and light and heavy fuel oil. Those suppliers must record the amount of fuel purchased or consumed each year and, using government-provided emissions factors, calculate total emissions associated with that fuel. They must then surrender an equivalent number of emission allowances to the government before the annual compliance period the following year. Biofuels used in the transport sector are not covered by the ETS [8].

## 3 The Western Climate Initiative (WCI)

The WCI is a regional ETS among subnational entities, namely the US state of California and the Canadian province of Quebec. In 2018, it also included the province of Ontario.[1] The jurisdictions involved differ from the emissions profile of North America overall in that a greater share of their greenhouse gas output comes from the transport sector. In Quebec's case, this is because the province's power is generated almost entirely by hydroelectric facilities, which emit virtually no greenhouse gases during routine operation—an ETS covering stationary sources like power plants and factories would thus accomplish little in terms of emission reduction for the

---

[1] Ontario implemented a provincial ETS in 2017, which linked to the WCI starting in January 2018. In June of that year, conservative Doug Ford was elected premier of the province—he had run on a platform promising to abolish the ETS, and withdrew the province from the WCI as one of his first acts in office.

province. Similarly, in California, which is the world's fifth largest economy, emissions from motor vehicles make up a much larger (and in 2008, when the ETS was being developed, faster-growing) portion of the state's greenhouse gas output than in most regions to which an ETS had previously been applied. While power and industry combined accounted for roughly 40% of California's emissions at the time the ETS was being developed, emissions from the transport sector and from residential/commercial fuel use accounted for nearly half of the state's GHG output (transport alone for roughly 40% by 2014 estimates) [9].

Thus, the WCI, which began with a pilot phase involving only stationary sources in 2013, was designed to include the transportation sector as of 2015. Similar to New Zealand, the point of obligation in the WCI is where the fossil fuel whose emissions need to be accounted for *enters commerce* in California or Quebec. This is the so-called terminal rack where oil and gas are physically transferred [10]. The owners of these facilities (oil and gas companies) pass the costs of allowances reflecting the embedded greenhouse gases to the consumer: they charge more for their fuel to cover the cost of allowances. Allowance prices in the WCI have ranged from USD$11.50 to nearly $19.00 in mid-2020 [11].

# 4 The Transportation and Climate Initiative (TCI)

The TCI is not an existing ETS, but rather a group of 13 north-eastern and mid-Atlantic jurisdictions in the USA that have been working together to reduce transport emissions in their region for the past half-decade. The collaboration has considered several policies designed to cut carbon dioxide from the transport sector, which makes up over 40% of the region's greenhouse gas emissions [12]. In this context, the jurisdictions have explored market-based policies including a so-called cap-and-invest programme that incorporates most features of an ETS combined with an upstream point of regulation. If the programme is launched,[2] it will constitute an ETS for the transportation sector, specifically for motor vehicles.

The participating jurisdictions are the District of Colombia (Washington DC, which is not a state) and the states Connecticut, Delaware, Maine, Maryland, Massachusetts, New Hampshire, New Jersey, New York, Pennsylvania, Rhode Island, Vermont, and Virginia. Together, they comprise 72 million people with 52 million

---

[2]The TCI is an ongoing public process, the most recent iteration of which involved circulation of a draft Memorandum of Understanding (MoU) that would (if signed by the authorities in the relevant jurisdictions) launch the cap-and-invest programme among those jurisdictions. A public comment period on the draft MoU ended in March 2020, with a final version to be agreed in summer 2020. If adopted by all jurisdictions in 2021, the TCI could enter into force January 2022—estimates of reduction trajectories are pegged to that year as the 'start date'. The declining emissions cap would be set for ten years through the end of 2031. However, the Covid-19 pandemic has slowed the timeline for adoption of the final MoU to at least Q4 2020, such that the respective regulatory agencies would have but a year to adopt the programme—stakeholders thus increasingly find a programme start by January 2022 unlikely.

registered vehicles and generate US \$5.3 trillion in GDP [13]. That is roughly one-fifth of the US total for all three metrics.

As explained above, an ETS typically involves a preset overall cap on emissions that declines over time, often broken into various compliance periods. The collective emissions of all the covered entities may not exceed the annual cap, as there are only enough allowances for the amount of emissions the cap allows for each year. TCI regulators have not decided on the cap yet, but have modelled policy scenarios that represent caps requiring 20, 22, and 25% carbon dioxide reductions, respectively, in the region's transport emissions during the 2022–2032 timeframe. While those cuts seem ambitious, forecasts of the region's emission trajectory *without* an ETS over the same timeframe project a 19% decrease in carbon dioxide from the transportation sector. Accordingly, the targets constitute only one, three, and six percent more pollution reduction than would occur under a 'business-as-usual' scenario for the region. The reference scenario assumes relatively fast penetration of electric vehicle infrastructure given the subsidies in the states involved, as well as continuation of US federal vehicle emissions standards [12].[3]

Unlike the New Zealand and WCI examples, the TCI would cover only the fossil fuel components of motor gasoline (petrol) and on-road diesel fuel destined for final sale or consumption in a TCI jurisdiction—not any other type of fuel. Covered vehicles include not only light-duty cars and trucks, but also commercial light trucks, freight trucks, and buses. The cap thus does not cover, e.g. rail transport. Like the WCI, the point where a compliance obligation applies is upon removal from a fuel storage facility or 'terminal rack' in the participating jurisdiction [12]. The companies that buy and sell fuel at this point in the distribution chain must surrender allowances for the emissions that fuel causes when combusted in vehicles.

## 5 The Carbon Offsetting and Reduction Scheme for International Aviation (CORSIA)

Greenhouse gas emissions from air travel, which currently account for three percent of global greenhouse gas output, are growing rapidly as flying becomes more affordable to ever-wider sections of the global population [14]. Whereas individual countries can implement measures at the national level to incentivise reduction of emissions in *domestic* air travel, *international* aviation (flights between countries) makes up the lion's share of global aviation emissions and is growing fastest. Despite its global nature, international aviation is not subject to the Paris Agreement under the UN Framework Convention on Climate Change. Instead, the body governing

---

[3]When sensitivity analysis was applied to the reference case, it revealed high sensitivity to the federal fuel economy standards US president Donald J Trump was dismantling, and to lower oil prices. A scenario in which federal fuel economy standards are rolled back and oil prices remain low thus yields only six percent reductions in transport emissions during 2022–2032, meaning emission reductions of 20, 22, or 25% in the 2022–2032 timeframe go well beyond 'business as usual'.

international aviation (the International Civil Aviation Organisation under the UN, ICAO), which is made up of member countries represented in an Assembly and a Council, resolved in 2016 [15] to address the sector's emissions growth through offsetting.

Under the Carbon Offsetting and Reduction Scheme for International Aviation (CORSIA) resolved in that year, average global greenhouse gas emissions from international aviation in the years 2019–2020 serve as a baseline[4]—starting in 2021, emissions above that level must be *offset* (also sometimes described as 'compensated' or 'neutralised') by air carriers. Offsetting in this case means purchasing emission reduction units from projects that cut greenhouse gases relative to a business as usual scenario. The project's developers quantify this reduction, have it verified and certified, and can then sell the unit as an 'offset' to buyers who use it to 'neutralise' their carbon emissions.

Although not an ETS per se, this system involves tradable units that put a price on emitting greenhouse gases and thus provide a financial incentive to cut carbon dioxide in the aviation sector. The degree to which air carriers are incentivised to do so depends on the price of offsets, which is in turn determined by demand for them. The more countries participate in CORSIA, the more routes are covered by the offsetting requirement [16]; higher participation rates, in turn, make for higher offset demand and thus more climate change mitigation activity. So far, 85 countries representing over three-fourths of international aviation activity have formally declared they will participate in CORSIA from the pilot phase starting January 2021 [17]. However, large nations that have significant and fast-growing aviation sectors (including Brazil, China, India, and Russia) have not committed to participating yet [18].

ICAO has put in place a Technical Advisory Body to recommend which standards of projects should be eligible under CORSIA [19]—those approved so far[5] involve programmes that, e.g., reduce fuel use in rural areas through more efficient cook-stoves/appliances, replace fossil-fuelled electricity generation with solar, wind, or small-scale hydro, or tree planting/reforestation initiatives and programmes. Using a formula to calculate their share of the amount by which the sector as a whole has

---

[4]Due to the global Covid-19 pandemic's extreme effect on amount of international air travel, 2020 emissions from flights between countries will be orders of magnitude lower than predicted when CORSIA's rules were established. This means that the baseline from which emissions growth is to be calculated (average of 2019–2020) will be lower, such that air carriers will need to buy *more* offsets in an optimistic aviation sector recovery scenario than they had expected before the pandemic. The International Air Transport Association (the airline industry lobby group IATA) and many national governments are calling for ICAO to change the rules of CORSIA to make the baseline emissions only 2019, not 2020. At the time of writing, ICAO's Assembly meeting was taking place and this issue was being discussed. A decision to change the baseline will impact the demand for offset units from the aviation sector, and therefore the viability of offset projects: lower demand makes for lower offset prices, meaning fewer projects in renewable energy, reforestation, and so forth, will be financially feasible for developers.

[5]Upon the TAB's recommendations, ICAO's Council, at its March 2020 meeting, approved six offset standards as eligible for use by air carriers under CORSIA. Not all units from these programmes are eligible, but the organisations that set the standards for what constitutes an offset unit have been 'pre-approved'.

exceeded the baseline each year, air carriers must submit an offset unit for every metric tonne carbon dioxide equivalent of emissions growth relative to the baseline. Carriers have several years to assess their offsetting obligations, having to comply with emissions in the 2021–2023 pilot phase only in 2025. Only emissions from flights between countries that have agreed to participate in CORSIA from 2021 will be subject to this offsetting requirement by the respective airline. Known as the scheme's first voluntary compliance phase, this period is followed by more voluntary phases until CORSIA becomes mandatory for all countries in 2027 [20].

## 6    How Good Are ETS at Reducing Transportation Emissions?

The foregoing sections have explained the way different ETS have included emissions from transportation without weighing in on how *effective* that practice has proven for reducing those emissions—especially compared to other policies aimed at cutting the greenhouse gas output of transportation, such as direct taxation of vehicle fuels, municipal planning that reduces the need for individual vehicles, or outright bans on some types of transport. The programmes in question have not been around long enough to draw major conclusions on this question, and for New Zealand and the WCI, the inherent interaction between transportation and other sectors makes it impossible to tease out effects on transport specifically. Still, evidence from analyses of the latter two ETS by their own governments points to a relatively *low* impact of the ETS price incentive on transport emission reduction compared to other policies in the sector.

Reviews of the NZ ETS suggest that it has had only a minor impact on the country's emissions, due to low allowance prices [21]. For transport emissions specifically, the small price impact of the ETS has not evoked enough behavioural change to cut vehicle emissions: at prices of NZ$20 per tonne, a review of the ETS by the New Zealand Productivity Commission [22] reports that the emissions component of fuel prices are NZ$0.05 per litre of petrol or approximately 2.5% of the pump price. The Commission concludes that only much higher NZ ETS prices would increase behavioural change in the transport sector to a degree that emissions decrease can be attributed to the ETS, such that complementary measures will be required to reduce transport emissions, which grew 12% between 2010 and 2017 [23].

The situation is more complex in the Western Climate Initiative, as that ETS covers two different jurisdictions of vastly different size in different countries. Regulators in both jurisdictions did not intend for the ETS to be the main tool in their respective emission reduction toolbox. Especially in California, several other policies contribute to the transportation sector's annual greenhouse gas emission reductions—including a unique low carbon fuel standard [24] and standards for new and light-duty vehicles approved in 2004 that required automakers to achieve fleet-wide fuel economy improvements in the model years 2009–2016 [25]. However, given that direct taxes on transportation fuels have been considered politically dead in the water in North America (especially the USA) even though they provide a more direct incentive to the consumer to use less fuel [26], including transport emissions in the cap-and-trade programme was a way for California and Quebec policymakers to implement a price signal on fuel consumption without resorting to a fuel tax—even if this only made for a very small fuel price increase and thus minimal expected behavioural change.

The TCI's own estimation of its ability to reduce the region's motor vehicle emissions beyond the reductions expected anyway over the next decade—the aforementioned one to six percent—shows that other policies (such as subsidies for electric vehicles and fuel efficiency standards incorporated into the reference scenario) are assumed to have a bigger impact when it comes to transport sector carbon dioxide cuts. For TCI jurisdictions, an incentive to use a market-based policy in addition to other policy measures is that 10 member states are part of a regional ETS that covers only the electricity sector (the Regional Greenhouse Gas Initiative, RGGI). An eventual link between the TCI and RGGI would make for a regional ETS with wider scope, such that covered entities can take advantage of a wider range of carbon abatement costs among a larger number of covered entities.

As for CORSIA, the programme does not even seek to reduce emissions from international aviation, only to compensate for their expected exponential *increase* by financing emission reduction in other sectors. This at least facilitates advances in emission reduction practices outside of transport and fosters sustainable development projects. At the same time, by relying on offset methodologies that rely on a counterfactual assumption and robust emissions accounting, it brings about a non-trivial risk of fraudulent or manipulative behaviour that could ultimately result in a net emissions increase and undermine the environmental objectives of the CORSIA programme [27].

Incorporating transportation emissions into an ETS can have desirable *distributional effects*, as the allocation of allowances makes for a revenue recycling opportunity. Transport sector emitters covered upstream (fuel suppliers) are the entities needing the majority of the allowances issued by California each year, for instance. As opposed to emitters in the industry sector, which are allocated certain percentages of their compliance obligation for free, fuel suppliers pay for every allowance they need to meet their compliance obligations. This massive state revenue from selling allowances to covered entities in the transport sector (mainly large oil and gas companies) is invested in programmes at the local level, effectively 'redistributing' allowance proceeds from corporate entities to public and community projects such as greening urban areas or financing public transport that have a (dispersed) overall emission reduction effect. To enhance these distributional effects, the California legislature required at least one-fourth of the annual proceeds from allowance sales to be spent on projects specifically benefitting disadvantaged communities [28].

# References

1. R. Schmalensee, R.N. Stavins, The SO$_2$ allowance trading system: the ironic history of a grand policy experiment. J. Econ. Perspect. **27**, 103–122 (2013)
2. J. Opschoor, H. Vos, *Economic Instruments for Environmental Protection*. OECD Publishing (1989)
3. International Climate Action Partnership, 'New Zealand Emissions Trading Scheme'. ETS Detailed Information. International Carbon Action Partnership (2020), https://icapcarbonaction.com/en/?option=com_etsmap&task=export&format=pdf&layout=list&systems%5B%5D=48. Accessed 23 June 2020
4. New Zealand Ministry of Business, Innovation and Employment, New Zealand Energy Sector Greenhouse Gas Emissions: Annual Emissions Data Table, (2020) https://www.mbie.govt.nz/assets/Data-Files/Energy/annual-emissions-data-table.xlsx. Accessed 23 June 2020
5. Ministry for the Environment, New Zealand's Fourth Biennial Report under the United Nations Framework Convention on Climate Change (2019), https://www.mfe.govt.nz/publications/climate-change/new-zealands-fourth-biennial-report-under-united-nations-framework. Accessed 23 June 2020.
6. C. Leining, S. Kerr, A Guide to the New Zealand Emissions Trading Scheme. Motu Economics and Public Policy Research (2018), https://motu.nz/assets/Documents/our-work/environment-and-agriculture/climate-change-mitigation/emissions-trading/ETS-Explanation-August-2018.pdf. Accessed 23 June 2020
7. New Zealand Environmental Protection Agency, 2018 New Zealand Emissions Trading Scheme Report (2019), https://www.epa.govt.nz/assets/Uploads/Documents/Emissions-Trading-Scheme/Reports/Annual-Reports/b450975461/2018-ETS-Annual-Report.pdf. Accessed 23 June 2020
8. New Zealand Environmental Protection Agency, Industries in the Emissions Trading Scheme (2020), https://epa.govt.nz/industry-areas/emissions-trading-scheme/industries-in-the-emissions-trading-scheme/. Accessed 23 June 2020
9. California Air Resources Board, California Greenhouse Gas Emissions for 2000 to 2017—Trends of Emissions and Other Indicators (2019), https://ww2.arb.ca.gov/ghg-inventory-data. Accessed Mar 2020
10. California Air Resources Board, 'Information for Entities That Take Delivery of Fuel for Fuels Phased into the Cap-and-Trade Program Beginning on January 1, 2015.' Archived guidance document for fuel distributors (2014), https://ww2.arb.ca.gov/mrr-data. Accessed 30 Mar 2020
11. Intercontinental Exchange, Pricing history of the annual benchmark contract (the CCA future for delivery in that year) (2020) https://www.theice.com/products/53169040/
12. Transportation and Climate Initiative, Draft Memorandum of Understanding of the Transportation and Climate Initiative—For Stakeholder Input. Published 17 December 2019, https://www.transportationandclimate.org/main-menu/tcis-regional-policy-design-process-2019. Accessed 30 Apr 2020
13. Transportation and Climate Initiative, Public webinar, 17 December 2019: 'Cap-and-Invest Modelling Results.' Slide 4/49 (2019), https://www.transportationandclimate.org/main-menu/tcis-regional-policy-design-process-2019#Framework. Accessed 30 Apr 2020
14. European Union Aviation Safety Agency, European Aviation Environmental Report, (2019) https://www.easa.europa.eu/eaer/. Accessed 23 June 2020
15. International Civil Aviation Organisation, Assembly Resolution 39–3 (2016), https://www.icao.int/Meetings/a39/Documents/Resolutions/a39_res_prov_en.pdf. Accessed 23 June 2020
16. International Air Transport Association, COVID-19 and CORSIA: stabilizing net CO$_2$ at 2019 "pre-crisis" levels. Position paper 19 May, 2020, https://www.iata.org/contentassets/fb74546000050c48089597a3ef1b9fe7a8/covid19-and-corsia-baseline-190520.pdf. Accessed 23 June 2020
17. ICAO, Online guidance document about CORSIA 'What is CORSIA and how does it work?' (2020), https://www.icao.int/environmental-protection/Pages/A39_CORSIA_FAQ2.aspx. Accessed 23 June 2020

18. International Civil Aviation Organisation (ICAO) (2020) Continued progress toward implementation of ICAO's Carbon Offsetting and Reduction Scheme for International Aviation. Online press release from 22 June 2020, https://www.icao.int/Newsroom/Pages/Continued-progress-toward-implementation-of-ICAOs-Carbon-Offsetting-and-Reduction-Scheme-for-International-Aviation-.aspx. Accessed 23 June 2020
19. ICAO, Terms of Reference for the Technical Advisory Body (2019) https://www.icao.int/env ironmental-protection/CORSIA/Documents/TAB/TOR%20of%20TAB.pdf. Accessed 23 June 2020
20. ICAO, Online guidance document about CORSIA 'What is CORSIA and how does it work?' https://www.icao.int/environmental-protection/Pages/A39_CORSIA_FAQ2.aspx. Accessed 23 June 2020
21. C. Leining, S. Kerr, Lessons Learned from the New Zealand Emissions Trading Scheme. Motu Working Paper 16–06. Motu Economics and Public Policy. https://motu-www.motu.org.nz/wpapers/16_06.pdf. Accessed 30 Mar 2020
22. New Zealand Productivity Commission, Low-Emissions Economy: Final Report (2018), www.productivity.govt.nz/low-emissions. Accessed 30 Mar 2020
23. New Zealand Ministry for the Environment, New Zealand's Fourth Biennial Report under the United Nations Framework Convention on Climate Change (2019), https://www.mfe.govt.nz/publications/climate-change/new-zealands-fourth-biennial-report-under-united-nations-framework. Accessed 30 Mar 2020
24. California Air Resources Board, Presentation 'LCFS basics.' (2020), https://ww2.arb.ca.gov/resources/documents/lcfs-basics. Accessed 30 Mar 2020
25. US Environmental Protection Agency and Dept. of Transportation, Light-Duty Vehicle Greenhouse Gas Emission Standards and Corporate Average Fuel Economy Standards; Final Rule (2010), https://www.nhtsa.gov/corporate-average-fuel-economy/documents-associated-mys-2012-2016-rulemaking. Accessed 2 Apr 2020
26. P. DeCicco, L. Truelove, Carbon taxes and the affordability of gasoline. Report from the University of Michigan Energy Survey. Ann Arbor, MI: University of Michigan Energy Institute, September 2017
27. T.L. Brewer, M.A. Mehling, Transparency of climate change policies, markets and corporate practices, in *The Oxford Handbook of Economic and Institutional Transparency*. ed. by J. Forssbaeck, L. Oxelheim (Oxford University Press, Oxford, 2014), pp. 179–197
28. California Climate Investments, Annual report to the legislature (2020), https://www.caclim ateinvestments.ca.gov/annual-report. Accessed 23 June 2020

# The Future of Transportation Emission Issues

**Thomas Brewer**

**Abstract** The coronavirus pandemic has already had significant short-term consequences for transportation emissions, particularly the much lower levels of emissions resulting from the declines in the use of all types of transportation (except bicycles and walking). How rapidly and how far the levels of emissions will rise during the next several years is an important question. The answer is substantially dependent on macro-economic conditions, as well as government policies, firms' decisions, and public opinion. One possibility is that the deterioration of firms' financial positions will inhibit their investments in energy efficiency equipment and other emission-reducing measures. At the same time, there are many opportunities to improve transportation systems' infrastructure through government economic stimulus programs that can reduce emissions. Some projects are already in progress long before the pandemic and recession are over: installing electric recharging stations for motor vehicles, expanding high-speed rail systems, and developing seaport infrastructure to reduce the use of ships' auxiliary diesel engines. There may also be a shift in public opinion, industry practices, and government policies in reaction to the extent to which soot from transportation, and other sectors, has contributed to the pandemic death rate by causing weak lungs and other unhealthy preconditions that make people more vulnerable to the coronavirus and to the extent to which soot has been a carrier of the virus from local 'hot spots' to wider regional, national, and even international areas.

## 1 Introduction

This chapter briefly reviews a broad range of scenarios for transportation emissions in the aftermath of the coronavirus pandemic and related economic recession. The chapter is more speculative and less conclusive than the other chapters. It is intended to provoke ideas about what the future may be like, though it also reports the results

T. Brewer (✉)
Georgetown University, Washington, DC, USA
e-mail: brewert@georgetown.edu

of others' new modeling exercises with quantified scenarios. During the balance of 2021 and beyond, there will surely be many model-based scenarios of previously unimagined paths for transportation systems and their emissions.

The diverse speculations from business leaders and government officials in the box below are useful reminders of the uncertainties about the context in which transportation-specific issues will be addressed.

---

### Box: Images of the Post-pandemic Future [1]

- While the virus has emboldened nationalist politicians, it's also sparking hope among some policy wonks that global leaders will be reminded how important it is to work together. More collaboration would be good news for some of the world's most daunting challenges, everything from fighting climate change to strengthening supply chains. [Bloomberg introduction]
- The pandemic stresses us. It reminds us that we are connected. It reminds us that global supply chains, personal relationships, everything is connected. We don't win alone. We don't even win as small groups or nations. We win more broadly than that. We start to think a little more broadly about problems like global warming,…. [Stanley McChrystal]
- …[W]e're going to become more and more nationalistic. We're becoming more isolated and less globalist. It comes at a time when we need cooperation for the climate. [Alan Patricof]
- There is a compelling case for international cooperation and coordination in addressing the crisis and keeping its devastation to a minimum. If that coordination takes place and is successful, it could provide the basis for the next stage of international cooperation, integration, and economic growth. [Anne Krueger]
- The pandemic has proved the extent to which we can still be effective and create value away from our office desks. For a country like Japan, where there is still a tendency to measure performance by hours spent at the office, I believe there could be important implications for gender diversity in the workplace.… closing the gender employment gap could lift Japan's GDP by 10%,…. [Kathy Matsui]
- Essentially the coronavirus will make the world look more like China, in terms of the state's involvement in private-sector activities. It is the U.S. converging to China, not the other way around. [Stephen Jen]
- I can't imagine companies are going to go back to spending as much on business travel. [Susan Lyne]
- The biggest change will be how businesses look at the supply-chain issue. That's the 1,000-lb gorilla. Will companies that are dependent upon China for essential parts for their businesses move production out of China or at least second-source out of China? [Jim Chanos]

- We're all going to be permanently scarred by having lived through this pandemic. How soon will anybody get on an airplane? How soon will anybody stay in a hotel? [Sam Zell]

## 2 Macro-economic Conditions

As of mid-2020, there was no doubt that the world economy in general and the transportation sector in particular were in the midst of an historically significant, unprecedented recession. The pandemic-related recession was more severe than the 2008–2009 financial recession. During the first half of 2020, forecasts of economic conditions in the world, of course, became increasingly negative over time. For instance, the IMF forecast in its April 2020 *World Economic Outlook* a −3.0% decline in world gross product in 2020; by its June Outlook, it was forecasting −4.9% for 2020 [2]. The transportation sector was being particularly hard hit already in reality. International trade in goods and services (important in both aviation and maritime shipping) was expected to have declined by 12% by the end of 2020.

Regardless of the forecasted or eventual real conditions in the world economy and major national economies within it, there was much discussion about what the substance of economic recovery programs should be.

## 3 IEA Sustainable Recovery Special Report

The International Energy Agency (IEA) in collaboration with the International Monetary Fund (IMF) published a special report in June 2020 that included a section on transportation in its wide-ranging analysis of options for including sustainable projects in governments' economic recovery programs [3]. The transportation projects were in three groups: consumer purchases of more efficient vehicles, urban infrastructure, and high-speed rail. The report estimated reductions in emissions and increases in employment associated with projects. The results of their analysis are summarized in Table 1.

## 4 Examples of Local Actions and Concerns

A Danish urban bus program announced in June 2020 is an example of a national-level initiative with local-level implementation [4]. The program includes cooperation agreements between the national Ministry of Transport and six of the largest municipalities in the country. A common element is a commitment that all new buses

**Table 1**  Illustrative sustainable transportation projects in IEA report [3]

| Projects | Emission reductions | Employment increases |
|---|---|---|
| *Consumer purchases of more efficient vehicles* | | |
| Use scrappage incentive program to replace old inefficient cars with new efficient ones | 40% reduction in lifecycle $CO_2$ emissions in most regions | [2008–2009 incentives in USA increased more fuel efficient new car purchases—and created or retained 40,000–120,000 jobs in USA] |
| Replace internal combustion engine (ICEs) cars with battery electric vehicles (BEVs)—and hybrid electric vehicles | [$CO_2$] emission reductions of 80% in EU, 60% in USA, 40% in China. NOx emission reductions of virtually 100% | Manufacturing engines with 200 components for BEVs, compared to 1400 in ICEs, reduces jobs by 20,000, but manufacturing batteries increases jobs by that much or more, depending on the location of production |
| *Urban infrastructure* | | |
| Install recharging infrastructure for electric cars, trucks, vans, buses, bikes, and scooters | NA | Electrification of urban bus systems would create 30% more construction jobs than similar level of investment in roads |
| *High-speed rail (HSR)* | | |
| Displace 20% of air flights in North America, 10% of flights in Europe, and almost 8% of flights in Asia–Pacific with high-speed rail | Reduce $CO_2$ equivalent emissions by 18 grammes per passenger kilometer [or by more than 90%] | Worldwide HSR projects of 32,000 km under construction employ 2.6 million construction workers |

will be zero-emission by a certain date (some in 2020) and that all diesel buses will be phased out by a certain date. In addition, there are individualized commitments by some cities. One might say the 'future is now'—at least in some countries and cities. The particulars are indicated in the box below.

---

**Box: Danish Cities' Commitments to Change Their Bus Fleets [4]**

- Copenhagen: All new buses will be zero-emission beginning in 2020, and all diesel buses will be converted to zero-emission by the end of 2025. The potential establishment of zero-emission zones will be explored with other ministries.
- Frederiksberg: All new buses will be zero-emission. Diesel buses will be phased out by 2025. Also, by 2025 all citizens of the city will be able to leave their electric cars at re-charging stations no more than 250 m from their homes. In 2030, all city vehicles will be $CO_2$ neutral.

- Aarhus: New city buses and automobiles will be zero-emission beginning in 2020. All new city road vehicles will run on $CO_2$ neutral propellants or be zero-emission beginning in 2025.
- Odense: New city buses and passenger cars will be zero-emission beginning in 2021. The city must be climate neutral in 2030.
- Vejle: All new city buses will be zero-emission beginning in 2021. In 2021, a plan for all city passenger cars and other vehicles will be developed.

Yet, in June 2020, there was also early cautionary evidence about scenarios of emission paths in the aftermath of the pandemic. A report based on extensive data collected by the European Environment Agency (EEA) [5] found that many cities' emissions of nitrogen dioxide had already 'rebounded' significantly from their pandemic lows. One—Budapest—already had levels by late June that exceeded levels from before the 'lockdown.' Many other major cities experienced substantial increases from their lows. Paris, for instance, had gone from 14 μg per cubic meter to 30, Brussels from 16 to 30, and Milan from 19 to 33. An underlying contributing factor to the rapid rebounds was an aversion to public transportation.

At the same time, there was increasing evidence that air pollution, including transportation particulate emissions, has worsened the coronavirus pandemic by weakening people's respiratory systems so that they are more vulnerable to the effects of the coronavirus [6]. Thus, local areas in the USA that experience relatively intense air pollution—as measured by levels of $PM_{2.5}$—have also experienced more serious consequences of the coronavirus pandemic. In short, black carbon and other pollutants that cause lung cancer and heart disease also thereby indirectly cause more serious effects from the pandemic.

## 5 The Short Term and the Long Term

Lest a focus on the short-term effects of the pandemic and recession lead one to imagine a long term all the way to 2050 without a resumption of economic growth and associated increases in emissions without significant technological and policy changes, it is useful to note the pre-pandemic-recession era projections of the OECD International Transport Forum reported in its 2019 *Transport Outlook* [7]. The results of its long-range scenario projections comparing 2050 with 2015 emission levels were that *global transportation $CO_2$ emissions would grow by 60%, assuming 'a current ambition scenario where current and announced mitigation policies are implemented.'* Of course, there would be significant variations among sectors in the projections: Total freight and non-urban passenger emissions would increase by 225%, while urban passenger transportation would decrease by 19%.

They also noted with much prescience that transportation policy 'must anticipate disruptions that originate outside the sector.' The disruptions could be emission-reducing or emission-increasing—or more likely a combination of them—of technological, economic, and political disruptions. Certainly, the period of 2019–2020, when this book was being written, was an era of unprecedented disruptions of many types—some of them reducing transportations emissions and some of them increasing such emissions.

Further, there have been significant variations among countries in the types and extent of the disruptions. Some countries have been especially active in developing and deploying new emission-reducing technologies. Some have explicitly incorporated emission-reducing projects in the transportation sectors of their economic recovery programs. Some local governments have adopted more restrictive policies on transportation emissions, even as their national governments were moving in the opposite direction.

Because many of the emissions have virtually immediate short-term public health effects, as well as long-term health and climate effects, the benefits of the limitations can be evident quickly. And because they are somewhat concentrated in and near large urban population centers, the limitations can affect large numbers of people. Since large urban population centers are also transportation hubs, the emissions of all of the transportation modes examined in this chapter and other chapters directly affect large portions of the population—in fact more than half in most countries. So there are significant, observable, clearly beneficial short-term effects of mitigation policies for the health of large numbers of people. There are also correspondingly large economic benefits from the reduced costs of health care. There is thus an economically, politically, and psychologically compelling case for improving local air quality by reducing transportation emissions.

In contrast, the long-term and widely distributed benefits of mitigating climate change pose other challenges. In particular, it is important to be aware of a basic distinction between greenhouse gas *emission levels* and *concentration levels*: It is the continually increasing *concentration* levels, not the monthly and annual fluctuations in *emission* levels, that are important over the long-run. For example, although global April 2020 $CO_2$ emissions were 17% lower than preceding average levels because of the economic impact of the coronavirus pandemic, *the concentration level increased to a new record high in May* [8]. Thus, when the monthly emissions were low compared with previous levels, the concentration level nevertheless broke a new record. The concentration level was more than 417 parts per million at the Mauna Loa Observatory—a new record. It is therefore important to be mindful of the long-term trends in concentration levels, even as we assess the short-term policy choices that governments make.

Fortunately, many of the policies that improve short-term, local air quality and thus prevent early deaths of local residents also improve long-term, global climate change trends.

# References

1. Bloomberg, *What Our Post-pandemic Future Looks Like.* 12 May (2020), https://www.bloomb erg.com. Accessed 12 May 2020. Used with permission
2. International Monetary Fund (IMF), *A Crisis Like No Other, An Uncertain Recovery* (2020), https://www.imf.org/en. Accessed 25 June 2020
3. International Energy Agency (IEA), *Sustainable Recovery: World Energy Outlook Special Report* (2020), https://www.iea.org/events/world-energy-outlook-special-report-on-sus tainable-recovery. Accessed 26 June 2020
4. Denmark, Ministry of Transport, *Announcement of Climate Cooperation Agreements with the Country's Largest Municipalities* (2020), www.trm.dk. Accessed 26 June 2020
5. L. Hook, S. Bernard, L. Abboud, Air pollution rebounds in Europe's cities as lockdowns ease. *Financial Times,* 24 June 2020. https://www.ft.com. Accessed 24 June 2020
6. X. Wu, R. Nethery, M. Sabath, D. Braun, F. Dominici, *Exposure to Air Pollution and Covid-19 Mortality in the United States: A Nationwide Cross-Sectional Study.* Harvard T.H. Chan School of Public Health, Department of Biostatistics (2020), https://projects.iq.harvard.edu/covid-pm. Accessed 27 June 2020
7. Organisation for Economic Cooperation and Development/International Trade Forum (OECD/ITF), *Transport Outlook* 2019 (OECD, Paris, 2019)
8. US National Oceanic and Atmospheric Administration (NOAA), Global Monitoring Laboratory, *Trends in Atmospheric Carbon Dioxide* (2020), https://www.esrl.noaa.gov/gmd/ccgg/trends/. Accessed 6 June 2020.

# Correction to: Road Transportation Emissions in India: Adopting a 'Hub' and 'Spoke' Approach Towards Electric-Driven Decarbonization

**Mahesh Sugathan**

**Correction to:**
**Chapter "Road Transportation Emissions in India: Adopting a 'Hub' and 'Spoke' Approach Towards Electric-Driven Decarbonization" in: T. Brewer (ed.),** *Transportation Air Pollutants*, **Springer Briefs in Applied Sciences and Technology,**
**https://doi.org/10.1007/978-3-030-59691-0_4**

The original version of the book was published with incorrect mail ID. The chapter Author's e-mail address have been updated as "msugathan.tradeprojects@gmail.com". The chapter and book have been updated with the changes.

---

The updated version of this chapter can be found at
https://doi.org/10.1007/978-3-030-59691-0_4

Printed in the United States
by Baker & Taylor Publisher Services